Complex Ball Quotients and Line
Arrangements in the Projective Plane

Complex Ball Quotients and Line Arrangements in the Projective Plane

Paula Tretkoff

With an appendix by Hans-Christoph Im Hof

Mathematical Notes 51

PRINCETON UNIVERSITY PRESS

PRINCETON AND OXFORD

Library of Congress Cataloging-in-Publication Data

Tretkoff, Paula, 1957–
Complex ball quotients and line arrangements in the projective plane /
Paula Tretkoff.
pages cm. – (Mathematical notes ; 51)
Includes bibliographical references and index.
ISBN 978-0-691-14477-1 (pbk. : alk. paper) 1. Curves, Elliptic.   2. Geometry,
Algebraic.   3. Projective planes.   4. Unit ball.   5. Riemann surfaces.   I. Title.
QA567.2.E44T74 2016
516.3′52–dc23   2015016120

British Library Cataloging-in-Publication Data is available

This book has been composed in Minion Pro

Printed on acid-free paper. ∞

press.princeton.edu

Typeset by S R Nova Pvt Ltd, Bangalore, India
Printed in the United States of America

1   3   5   7   9   10   8   6   4   2

To the memory of Friedrich Hirzebruch

Contents

Preface

This book is devoted to a study of quotients of the complex 2-ball yielding finite coverings of the projective plane branched along certain line arrangements. It is intended to be an introduction for graduate students and for researchers. We give a complete list of the known weighted line arrangements that can give rise to such ball quotients, and then we provide a justification for the existence of the ball quotients. The Miyaoka-Yau inequality for surfaces of general type, and its analogue for surfaces with an orbifold structure, plays a central role.

The book has its origins in a Nachdiplom course given by F. Hirzebruch at the ETH Zürich during the Spring of 1996. I (née Cohen) was at that time Directeur de Recherche au CNRS at the Université de Lille 1, and a guest of ETH Zürich. I attended the course, lectured on some of the material on hypergeometric functions, and made the original set of notes for all Hirzebruch's lectures. I also presented related material as a two-semester graduate course at Princeton University during the academic year 2001/2002 while I was Visiting Professor there. The ETH Nachdiplom course notes were subsequently developed and refined by F. Hirzebruch and me during regular visits to the Max Planck Institute in Bonn. I thank ETH Zürich, MPI Bonn, and Princeton University for their support during the preparation of this book.

After working on the book together for some years, F. Hirzebruch and I decided even more additional material seemed desirable and, at that point, F. Hirzebruch asked me to complete the book under my own name. The book entitled *Geradenkonfigurationen und Algebraische Flächen*, Vieweg, 1987, by G. Barthel, F. Hirzebruch, and T. Höfer, served as a valuable resource and reference. This book is in German, and we have mined some of its contents as needed. The present book assumes less background than the earlier book, and we revisit in more detail several of its important subjects. We have also added material not found in that book. For example, we present a more organized list of the possible weights on line arrangements that yield finite covers that are ball quotients, and we devote a whole chapter, rather than just a few pages, to the question of the existence of such ball quotients. Also, we include material on hypergeometric functions.

The book is dedicated to the memory of F. Hirzebruch. I am grateful beyond words for the time and attention he paid to this project amid the many commitments of his important and busy life.

Complex Ball Quotients and Line
Arrangements in the Projective Plane

Introduction

Historical precedent for the results in this book can be found in the theory of Riemann surfaces. Every compact Riemann surface of genus $g \geq 2$ has representations both as a plane algebraic curve, and so as a branched covering of the complex projective line, and as a quotient of the complex 1-ball, or unit disk, by a freely acting cocompact discrete subgroup of the automorphisms of the 1-ball. The latter result is a direct consequence of the uniformization theorem, which states that the simply-connected Riemann surfaces are the Riemann sphere, the complex plane, and the complex 1-ball.

In complex dimension 2, Hirzebruch (1956) and Yau (1978) showed that the smooth compact connected complex algebraic surfaces, representable as quotients of the complex 2-ball B_2 by freely acting cocompact discrete subgroups of the automorphisms of the 2-ball, are precisely the surfaces of general type whose Chern numbers satisfy $c_1^2 = 3c_2$. Here, c_2 is the Euler characteristic and c_1^2 is the self-intersection number of the canonical divisor. Accordingly, the *proportionality deviation* of a complex surface is defined by the expression $3c_2 - c_1^2$. This book examines the explicit computation of this proportionality deviation for finite covers of the complex projective plane ramified along certain line arrangements. Candidates for ball quotients among these finite covers arise by choosing weights on the line arrangements such that the proportionality deviation vanishes. We then show that these ball quotients actually exist.

The intention of F. Hirzebruch in the original notes was for the material to be presented in a nontechnical way, assuming a minimum of prerequisites, with definitions and results being introduced only as needed. The desire was for the reader to be exposed to the theory of complex surfaces—and of complex surfaces with an orbifold structure—through the examples provided by weighted line arrangements in the projective plane, with an emphasis on their finite covers which are ball quotients. There was no desire to develop a complete theory of surfaces and orbifolds although relevant references are provided, or to treat the latest developments, or to also survey results in higher dimensions. Instead, the goal was that, on reading the book, the student or researcher should be better equipped to go into more technical or more modern territory if interested. It was felt that the topic of the notes was important enough historically to warrant their appearance. F. Hirzebruch wanted the style of the original notes to be like that of his series of lectures, and for them to remain readable and conversational; and, even though a lot of material has been added since by me, I have aimed to retain those qualities.

The plan of the book is as follows. In Chapter 1, we collect in one place the main prerequisites from topology and differential geometry needed for subsequent chapters, although, for convenience, pieces of this material are at times repeated or expanded upon later in the book. Even if the reader is familiar with these topics, it is worth his or her while to leaf through Chapter 1 to see what notions and notations are used later.

In Chapter 2, we apply some of the material from Chapter 1 to Riemann surfaces—which are of real dimension 2—by way of a historical and conceptual motivation for the material on complex surfaces. We also discuss in this chapter the classical Gauss hypergeometric functions of one complex variable and the triangle groups that serve as their monodromy groups. When these triangle groups are hyperbolic, they act on the unit disk, that is, the complex 1-ball. Infinitely many such groups act discontinuously, and their torsion-free normal subgroups of finite index define quotients of the complex 1-ball that are finite coverings of the projective line branched over three points.

In Chapter 3, we study complex surfaces and their coverings branched along divisors, that is, subvarieties of codimension 1. In particular, we discuss coverings branched over transversally intersecting divisors. Applying this to line arrangements in the complex projective plane, we first blow up the projective plane at non-transverse intersection points, that is, at those points of the arrangement where more than two lines intersect. These points are called singular points of the arrangement. This results in a complex surface and transversely intersecting divisors that contain the proper transforms of the original lines. Next, we introduce the group of divisor classes, their intersection numbers, and the canonical divisor class. The Chern numbers c_1^2 and c_2, as well as the proportionality deviation, Prop $:= 3c_2 - c_1^2$, are then defined.

In Chapter 4, we give an overview of the rough classification of (smooth complex connected compact algebraic) surfaces. We present two approaches that, in dimension 2, give the Miyaoka-Yau inequality; $c_1^2 \leq 3c_2$, for surfaces of general type. We only make a few remarks about the first of these, due to Miyaoka, which uses algebraic geometry. The second, due to Aubin and Yau, uses analysis and differential geometry. Here we give more details, since we use the analogous approach to the Miyaoka-Yau inequality for surface orbifolds in Chapter 6. We also discuss why equality in the Miyaoka-Yau inequality characterizes surfaces of general type that are free quotients of the complex 2-ball B_2, that is, orbit spaces for the action of discrete cocompact subgroups of the automorphisms of B_2 that have no fixed points.

In Chapter 5 we arrive at the main topic of the book: the free 2-ball quotients arising as finite covers of the projective plane branched along line arrangements. The material of this chapter is self-contained and largely combinatorial. Let X be the surface obtained by blowing up the singular intersection points of a line arrangement in the complex projective plane. Let Y be a smooth compact complex surface given by a finite cover of X branched along the divisors on X defined by the lines of the arrangement, that is, by their proper transforms and by the blown-up points, also known as exceptional divisors. If Y is of general type with vanishing proportionality deviation Prop, then it is a free 2-ball quotient. In order to find such a Y, we use a formula for Prop due to T. Höfer. This formula expresses Prop as the sum of, first, a quadratic form evaluated at numbers related only to the ramification indices along the proper transforms of the lines, and, second, nonnegative contributions from each blown-up point. The quadratic form itself depends only on the original line arrangement. The contributions from the blown-up points vanish when we impose certain diophantine conditions on the choice of ramification indices. Next, we seek line arrangements and ramification indices that make the quadratic form vanish. Initially, we ask of the arrangements that they have equal ramification indices along each of the proper transforms of the original lines. For these arrangements, the number of intersection points on each line is $(k + 3)/3$, where k is the number of lines. We then list all known line arrangements with this property and restrict our attention to them. Next, we replace the ramification indices by weights that are positive or negative integers or infinity, and we list all possible weights that satisfy the diophantine conditions at the blown-up points and also annihilate the quadratic form. Finally, we enumerate all possibilities for the assigned weights of the arrangements, under the assumption that divisors of negative or infinite weight on the blown-up line arrangements do not intersect. When the weight of a divisor on a blown-up line arrangement is negative, the curves above it on Y are exceptional and can be blown down. Blowing down all curves on Y arising from such negative weights, we obtain a smooth surface Y'. When the weight of a divisor on a blown-up line arrangement is infinite, the curves above it on Y are elliptic. Letting C be the union on Y' of all elliptic curves arising in this way, we derive the appropriate expression for the Prop of the possibly non-compact smooth surface $Y' \setminus C$. It is known that we cannot have rational or elliptic curves on a free 2-ball quotient, so this construction is quite natural. For the line arrangements of Chapter 5, we give the complete list of weights such that the Prop of $Y' \setminus C$ vanishes, meaning that $Y' \setminus C$ is a free 2-ball quotient. These weights are all presented in a series of tables in §5.7 at the end of Chapter 5, except for a few extra cases that are listed in §5.6.1. Throughout this chapter, we assume that the finite covers of the line arrangements with vanishing proportionality actually exist. The expression for Prop of $Y' \setminus C$ depends only on the original weighted line arrangement, so we can work with this assumption; but we still need to show that there are such covers.

In Chapter 6 we justify the existence assumption of Chapter 5. Let X be the blow-up of the projective plane at the non-transverse intersection points of a line arrangement. Assign weights—allowed to be positive or negative integers or infinity, denoted by n_i for the proper transforms D_i of the lines, and denoted by m_j for the exceptional divisors E_j given by the blown-up points—in such a way that the divisors with negative or infinite weights are distinct. Let X' be the possibly singular surface obtained by blowing down the D_i and E_j with negative weight, and X'' the possibly singular surface obtained by contracting the images of the D_i and E_j on X' with infinite weight. Let D_i'' be the image in X'' of D_i with weight $n_i > 0$, and E_j'' the image in X'' of E_j with weight $m_j > 0$. The central question of Chapter 6 is the following. When is X'' a (possibly compactified by points) ball quotient $X'' = \overline{\Gamma \backslash B_2}$ for a discrete subgroup Γ of the automorphisms of the ball with natural map $B_2 \to X''$ ramified of order n_i along D_i'' and m_j along E_j''? Any normal subgroup Γ' of Γ of finite index N in Γ, acting on B_2 without fixed points, gives rise to a free ball quotient $\Gamma' \backslash B_2 = Y' \setminus C$, as described in Chapter 5, that is a finite cover of $X_0'' = \Gamma \backslash B_2$ of degree N, ramified of order $n_i > 0$ along $D_i'' \cap X_0''$ and $m_j > 0$ along $E_j'' \cap X_0''$. Such normal subgroups exist due to the work of Borel and Selberg.

Therefore, we approach the proof of the existence of $Y' \setminus C$ by first showing the existence of Γ. The group Γ acts with fixed points, so, in answering this central question, the language of orbifolds is appropriate. Given the simplicity and explicit nature of the orbifolds we study, we use instead the related notion of b-space due to Kato. For the reader seeking a more sophisticated and general treatment, it is worth learning the basics of complex orbifolds that we exclude. Roughly speaking, we show that if all the diophantine conditions on the weights, derived in Chapter 5 to ensure that Prop vanishes, are satisfied, then there is equality in an orbifold, or b-space, version of the Miyaoka-Yau inequality, and, by arguments analogous to those of Chapter 4, the space X_0'' is of the form $\Gamma \backslash B_2$ and the map $B_2 \to X_0''$ has the desired ramification properties described above. To do this, we use the work of R. Kobayashi, S. Nakamura, and F. Sakai [83], which generalizes the work of Aubin-Yau on the Miyaoka-Yau inequality to surfaces with an orbifold structure.

There are both related and alternative approaches dating from about the same time, for example, the approach in the independent work of S. Y. Cheng and S. T. Yau [27], as well as more modern and more general approaches, such as in the work of A. Langer [91], [92], to mention just a few important instances. We choose to base our discussion on the work of R. Kobayashi, S. Nakamura, and F. Sakai, as it was part of the original notes made with F. Hirzebruch and it fits well with the presentation of the work of Aubin-Yau in Chapter 4 as well as with our approach to the material of Chapter 5. The main point of Chapter 6 is to flesh out the details of the orbifold version of the Miyaoka-Yau inequality for our particular situation of line arrangements in the projective plane.

Chapter 7 focuses on the complete quadrilateral line arrangement, and especially on its relationship with the space of regular points of the system of partial differential equations defining the Appell hypergeometric function. The latter is the two-variable analogue of the Gauss hypergeometric function discussed in Chapter 2. Building on work of E. Picard and of T. Terada, Deligne and Mostow established criteria for the monodromy group of this Appell function to act discontinuously on the complex 2-ball. This leads to examples of complex 2-ball quotients, determined by freely acting subgroups of finite index in these monodromy groups that are branched along the blown-up complete quadrilateral. Some of these monodromy groups provide rare examples of non-arithmetic groups acting discontinuously on irreducible bounded complex symmetric domains of dimension greater than 1.

The book concludes with two appendices. The first is by H.-C. Im Hof[1] and supplies a proof of Fenchel's Conjecture about the existence of torsion-free subgroups of finite index in finitely generated Fuchsian groups. The second concerns Kummer coverings of line arrangements in the projective plane.

I thank Marvin D. Tretkoff and Robert C. Gunning for their useful advice and unfailing encouragement while I wrote this book.

---

[1] Professor at the Institute of Mathematics of Basel University, Basel, Switzerland.

Chapter One

Topological Invariants and Differential Geometry

In this chapter, we compile some prerequisites from topology and differential geometry needed in later chapters. For the most part we do not provide proofs since there are many good references for this material, for example, [19], [49], [55], [112]. In §1.1, for a topological space X, we define singular homology and cohomology, as well as the Euler number $e(X)$. The Euler number is the topological invariant that we will encounter the most often in the subsequent chapters. For a complex surface X, it coincides with the second Chern number $c_2(X)$, as we shall see in Chapter 3 (we assume there that X is smooth compact connected algebraic). In that chapter, we also introduce for such a surface the first Chern number $c_1^2(X)$, which can be defined as the self-intersection number of the canonical divisor. Some generalities on the first Chern class $c_1(X)$ as well as necessary background on the canonical divisor are given in §1.4, although intersection theory for surfaces is only introduced in Chapter 3. The Miyaoka-Yau inequality for minimal smooth compact connected algebraic surfaces of general type, which is of deep importance for the material of this book, is derived in Chapter 4, and is the inequality $c_1^2(X) \leq 3c_2(X)$ relating the Chern numbers $c_1^2(X)$ and $c_2(X)$. In Chapter 6, we derive a version of this inequality for surfaces with an orbifold structure that are not necessarily compact (we touch on the non-compact situation also at the end of Chapter 4). When this inequality is an equality, X is a quotient of the complex two-dimensional ball B_2 by a discrete subgroup of the automorphisms of B_2, acting without fixed points in Chapter 4 and with fixed points in Chapter 6. For the summaries of the proofs of the Miyaoka-Yau inequalities in these chapters, we use techniques due to Aubin, S.-T. Yau, and R. Kobayashi coming from differential geometry and partial differential equations. Some of the differential geometry can be found in §1.4 and the rest is derived as needed in Chapters 4 and 6.

In this book, we discuss the Miyaoka-Yau inequality only for surfaces, as our interest is in weighted line arrangements in the complex projective plane. A suitably generalized Miyaoka-Yau inequality due to Aubin and Yau holds, for example, for compact Kähler manifolds of dimension n whose first Chern class vanishes or is negative, meaning it is represented by a real closed negative definite $(1, 1)$-form (see §1.4 for definitions). For a statement, see [20], [84], p. 323.

1.1 TOPOLOGICAL INVARIANTS

Let X be a topological space. We briefly recall the definition, using singular chains, of the singular homology groups $H_i(X, \mathbb{Z})$ with integer coefficients (see, for example, [19], [112]). Viewing $\mathbb{R}^n$ as the subset of $\mathbb{R}^{n+1}$ consisting of the vectors with $(n+1)$th coordinate equal to 0, we can consider the union $\mathbb{R}^\infty = \cup_{n \geq 1} \mathbb{R}^n$. Let $e_n, n \geq 1$ be the vector whose nth coordinate is 1 and whose other coordinates are 0, and let e_0 be the vector with all its coordinates 0. For $i \geq 0$, the standard simplex Δ_i of dimension i is given by the set

$$\Delta_i = \{\sum_{n=0}^{i} \lambda_n e_n \mid \lambda_n \geq 0, \sum_{n=0}^{i} \lambda_n = 1\}.$$

A singular i-dimensional simplex in X is a continuous map from Δ_i onto X. The singular i-chains $C_i(X)$ in X are the finite linear combinations, with integer coefficients, of the singular i-dimensional simplices. They form an abelian group.

For $i \geq 1$ and $k = 0, \ldots i$, we define the kth face map $\partial_i^k : \Delta_{i-1} \to \Delta_i$ by

$$\partial_i^k \left(\sum_{n=0}^{i-1} \lambda_n e_n \right) = \sum_{n=0}^{k-1} \lambda_n e_n + \sum_{n=k+1}^{i} \lambda_{n-1} e_n, \qquad k = 1, \ldots, i-1,$$

with

$$\partial_i^0 \left(\sum_{n=0}^{i-1} \lambda_n e_n \right) = \sum_{n=1}^{i} \lambda_{n-1} e_n, \qquad k = 0$$

and

$$\partial_i^i \left(\sum_{n=0}^{i-1} \lambda_n e_n \right) = \sum_{n=0}^{i-1} \lambda_n e_n, \qquad k = i.$$

The boundary operator

$$\partial : C_i(X) \longrightarrow C_{i-1}(X)$$

is defined as the alternating sum

$$\partial s = \sum_{k=0}^{i} (-1)^k s \circ \partial_i^k, \qquad s \in C_i(X).$$

It satisfies $\partial\partial = 0$, and so we have a differential complex with homology group

$$H_i(X, \mathbb{Z}) = \frac{\text{Ker}(\partial : C_i(X) \to C_{i-1}(X))}{\partial C_{i+1}(X)}.$$

This definition depends only on X. No triangulation or similar structure is needed. If X is triangulated, we can use the chain groups $c_*(X)$ consisting of finite linear combinations of the simplices of the triangulation; otherwise the definition is as before. It is known that the homology groups obtained in this way are the groups $H_i(X, \mathbb{Z})$ defined above. If the abelian group $H_i(X, \mathbb{Z})$ is finitely generated, then the rank of $H_i(X, \mathbb{Z})$ is called the ith Betti number, $b_i(X)$. The subgroup of elements of finite order is the torsion subgroup $T_i(X)$. If the space X admits a finite triangulation, then the $H_i(X, \mathbb{Z})$ are all finitely generated and trivial for i sufficiently large. If $H_i(X, \mathbb{Z})$ is finitely generated, then $H_i(X, \mathbb{Z}) \otimes \mathbb{R}$ is a real vector space of dimension $b_i(X)$, which is also denoted by $H_i(X, \mathbb{R})$. The homology groups are homotopy invariants.

The dual construction gives singular cohomology. A singular i-cochain on X is a linear functional on the $\mathbb{Z}$-module $C_i(X)$ of singular i-chains. The group of singular i-cochains is therefore $C^i(X) = \text{Hom}(C_i(X), \mathbb{Z})$. The coboundary operator is defined by $(d\omega)(s) = \omega(\partial s)$, $\omega \in C^*(X)$, and satisfies $dd = 0$. The graded group of singular cochains $C^*(X) = \oplus_i C^i(X)$ is therefore a differential complex whose homology is called the singular cohomology of X with integer coefficients. The ith cohomology group is denoted by $H^i(X, \mathbb{Z})$ (for more details, see [19]).

The Euler-Poincaré characteristic, which we also call the Euler number, is an important invariant of a topological space X, and is denoted by $e(X)$. Assume $H_i(X, \mathbb{Z})$ is finitely generated and trivial for i sufficiently large. Then

$$e(X) = \sum_i (-1)^i b_i(X) = \sum_i (-1)^i \dim H_i(X, \mathbb{R}). \qquad (1.1)$$

The number $e(X)$ is a topological invariant and a homotopy invariant. If the space has a finite triangulation, then

$$e(X) = \sum_i (-1)^i \text{rank}(c_i(X)).$$

In other words,

$$e(X) = \sum_i (-1)^i e_i,$$

where e_i is the number of i-dimensional simplices of the triangulation. This generalizes to cellular decompositions.

The Euler-Poincaré characteristic generalizes the notion of the cardinality of a finite set. Indeed, if X is a finite set, then $e(X) = |X|$. If X and Y are finite sets, then we have

$$|X \cup Y| = |X| + |Y| - |X \cap Y|,$$

which generalizes to the Mayer-Vietoris relation

$$e(X \cup Y) = e(X) + e(Y) - e(X \cap Y), \qquad (1.2)$$

which is certainly true if $X \cup Y$ admits a finite triangulation that induces triangulations on X, Y, and $X \cap Y$. In all our applications of the Mayer-Vietoris relation, this will be the case. Moreover, if X and X' are homotopically equivalent, then $e(X) = e(X')$. For products of topological spaces (admitting finite triangulations), we have the formula

$$e(X \times Y) = e(X)e(Y),$$

generalizing

$$|X \times Y| = |X| \cdot |Y|$$

for the cardinality of finite sets. These simple formulas allow us to compute the Euler-Poincaré characteristic (Euler number) of many of the spaces we shall encounter. We can also calculate the homology groups, but prefer to use the Mayer-Vietoris relation and the product formula. In fact, in many cases occurring in later chapters, it is easy to calculate the Euler number but very difficult to determine the Betti numbers. For more details and proofs of these facts, see [99].

A point P is a set with one element; hence $e(P) = 1$. An interval I is homotopic to a point so that

$$e(I) = 1.$$

As the circle S^1 is the union of two semicircles homeomorphic to the unit interval and intersecting at two points, we have

$$e(S^1) = e(I) + e(I) - 2 \cdot e(P) = 0.$$

On the other hand, the 2-sphere S^2 (also called the Riemann sphere) is the union of two hemispheres, each homotopic to a point and intersecting in a copy of S^1, so that

$$e(S^2) = e(P) + e(P) - e(S^1) = 2.$$

More generally, the n-sphere S^n satisfies

$$e(S^n) = 2, \text{ for } n \text{ even}, \qquad e(S^n) = 0, \text{ for } n \text{ odd}.$$

Example: In Chapter 2 we discuss Riemann surfaces (real dimension 2). The real manifold underlying a compact Riemann surface is homeomorphic to S^2 with g handles attached, for a certain number g depending only on the topology of the Riemann surface. Each handle is homeomorphic to $I \times S^1$, where I is the unit interval and S^1 the unit circle. The invariant g is called the *genus* of the Riemann surface. From S^2 we remove g pairs of open disks (the closed disks are supposed to be disjoint, each being a disk with respect to a local coordinate system). By Mayer-Vietoris, the resulting space X has Euler number $2 - 2g$. The Euler number of the Riemann surface F_g of genus g also equals $2 - 2g$, because the handles have Euler number 0 and intersect X in circles. (Attaching a handle means identifying the two boundary circles of the handle with a pair of circles in X.) For F_g we have

$$e(F_g) = 2 - 2g = e - k + f,$$

where e is the number of vertices, k the number of edges, and f the number of faces of a triangulation, or more generally a cellular decomposition of F_g. For $g = 0$ (when $F_g = S^2$), this is the traditional formula of Euler (and perhaps Descartes). The classification of compact-oriented surfaces by their genus g is essentially due to A.F. Möbius (1863) and C. Jordan (1866). See the essay by R. Remmert and M. Schneider in [142].

1.2 FUNDAMENTAL GROUPS AND COVERING SPACES

In this book we encounter quotients of simply-connected complex spaces by groups of automorphisms of the spaces acting discontinuously, and often freely. For example, in Chapter 2 we use the fact that every Riemann surface is the quotient of one of the simply-connected Riemann surfaces, called its universal cover, by a subgroup acting freely discontinuously. This fact is a consequence of the theory of fundamental groups and covering spaces. Here, we give only a rough indication of some of the basics of this theory. More details can be found, for example, in [71], Chapter IX. Let X be a topological space, and x_0 a fixed point of X. We say X is path-connected if there is a path between any two points of X. Suppose X is path-connected. The paths in X with base point x_0 are the continuous maps p from the interval $[0, 1]$ to X with $p(0) = x_0$. We can put the compact open topology on this space, with the basic open sets consisting of the paths p in X with base point x_0 such that $p(K) \subset U$ for a fixed compact set K in $[0, 1]$ and a fixed open set U in X.

The paths p with $p(0) = x_0$ and $p(1) = x_0$ are called the loops based at x_0 and have a natural composition law. For loops p_1, p_2, the composite $p_1 \circ p_2$ is obtained by first traversing p_1 and then traversing p_2. A homotopy between two paths p_1 and p_2 in X is a continuous map $F : [0, 1] \times [0, 1] \to X$ such that $F(t, s) = p_1(t)$ for $s = 0$ and $F(t, s) = p_2(t)$ for $s = 1$. This induces an equivalence relation between loops (based at x_0). The set of homotopy classes of loops (based at x_0), with the operation given by the composition of loops, is the fundamental group $\pi_1(X, x_0)$. Then, for any two base points x_0, y_0, the groups $\pi_1(X, x_0)$ and $\pi_1(X, y_0)$ are isomorphic. An isomorphism can be constructed by using the path from x_0 to y_0. Two such isomorphisms are related by inner automorphisms of the two groups. By definition, the path-connected space X is simply connected if and only if $\pi_1(X, x_0)$ is trivial.

Every path-connected space X, which satisfies some additional conditions, has a simply-connected universal covering space Y with a free action of $\pi_1(X, x_0)$ such that the orbit space is X. The additional conditions (see [71]) are locally path-connected and semi-locally simply connected. These certainly hold for manifolds and for triangulated spaces. For $x \in X$, we denote by $\Omega(X, x_0, x)$ the set of paths from x_0 to x. Elements α, β of $\Omega(X, x_0, x)$ define $\alpha \circ \beta^{-1}$ in $\Omega(X, x_0, x_0)$ and hence an element in $\pi_1(X, x_0)$. If this element is the identity element the elements α, β are called equivalent. The set of equivalence classes is denoted by Y_x. The space Y is the disjoint union of all these Y_x for $x \in X$. The natural projection $\pi : Y \to X$ has the Y_x as fibers, that is, $\pi^{-1}(x) = Y_x$. Fixing a path from x_0 to x, or rather a point of Y_x, defines a bijection of Y_x to $\pi_1(X, x_0)$, with the fixed point of Y_x going to the identity. Composing this fixed path with a path from x to an arbitrary point x' in a locally path-connected neighborhood U of x leads to a local trivialization $U \times \pi_1(X, x_0)$, where $\pi_1(X, x_0)$ carries the discrete topology, and Y carries a global topology which is locally of this form. The covering space Y is simply connected. If $\pi(y_0) = x_0$, then a loop α in Y from y_0 to y_0 projects under π to a loop $\pi(\alpha)$ from x_0 to x_0, which by our construction represents the identity element of $\pi_1(X, x_0)$. But a homotopy from $\pi(\alpha)$ to the constant loop can be lifted to a homotopy in Y deforming α to the constant loop.

If γ is an element of $\Omega(X, x_0, x)$ and α an element of $\Omega(X, x_0, x_0)$, then the composition $\alpha \circ \gamma$ of paths is an element of $\Omega(X, x_0, x)$. This induces an action of the group $\pi_1(X, x_0)$ on Y from the left. The group $\pi_1(X, x_0)$ operates freely discontinuously on Y, that is, for every point of Y there exists a neighborhood U such that $\alpha U \cap U = \phi$ for all $\alpha \in \pi_1(X, x_0)$, $\alpha \neq \mathrm{id}$.

As we shall recall in Chapter 2, if we take an arbitrary Riemann surface X (real dimension 2, compact or not), then its universal cover Y, often denoted by $\widetilde{X}$, is isomorphic to one of the simply-connected Riemann surfaces, and so to the complex plane, to the upper half plane, or to the complex projective line. Therefore, we may lift the complex structure on X to Y, with $\pi_1(X, x_0)$ operating by holomorphic automorphisms of Y. Thus, the Riemann surface

X is the quotient of one of the three simply-connected surfaces $\mathcal{D}$ by a freely discontinuous subgroup of $\mathrm{Aut}(\mathcal{D})$, which is therefore a discrete subgroup. This group is called the universal covering group. More generally, we can take a subgroup G of $\pi_1(X, x_0)$ and divide Y by G. Then $G\backslash Y$ is a covering space of X with $G\backslash\pi_1(X, x_0)$ as fiber. The space Y is also the universal cover of $G\backslash Y$, and G is the fundamental group of $G\backslash Y$. If G is a normal subgroup of $\pi_1(X, x_0)$, then $G\backslash\pi_1(X, x_0)$ acts on $G\backslash Y$ with X as orbit space.

Examples:

- The space $X = \mathbb{P}_1 \setminus \{0, 1, \infty\}$ occurs in Chapter 2 in our discussion of classical one-variable hypergeometric functions. Let $x_0 \in X$ be fixed. The fundamental group $\pi_1(X, x_0)$, which we also denote by $\pi_1(X)$, is generated by three loops, α_0, α_1, and α_∞, starting at x_0 and going once around 0, 1, and ∞, respectively, and subject to precisely one relation, given by

$$\alpha_0 \alpha_1 \alpha_\infty = \mathrm{Id}.$$

This relation expresses the fact that we can deform the composition of the three loops $\alpha_0, \alpha_1, \alpha_\infty$ over the Riemann sphere S^2, while avoiding $0, 1, \infty$, to get a closed loop not going around $0, 1, \infty$, which can then be deformed to 0. By the computations of §1.1, we know that $e(X) = -1$. The universal cover of X is $\mathcal{H}$ and the universal covering group is generated by three fractional linear transformations determined by three real 2×2 matrices, M_0, M_1, M_∞, with product the identity. This group is conjugate to the principal congruence subgroup $\Gamma[2]$, a Fuchsian triangle group of signature (∞, ∞, ∞) that we shall encounter again in Chapter 2.

- The space

$$Q = \{(x, y) \in \mathbb{P}_1^2 : x, y \neq 0, 1, \infty, \ x \neq y\}$$

appears in Chapter 7 in our discussion of the Appell hypergeometric function in two variables, which generalizes the classical one-variable hypergeometric function of Chapter 2. Let (x_0, y_0) be a fixed point of Q. For $j = 0, 1, \infty, x_0, y_0$, we can define a loop in Q which starts at (x_0, y_0) and either x or y travels in $\mathbb{P}_1 \setminus \{0, 1, \infty, x_0, y_0\}$ up to j, goes once around j in the positive direction, and does not wind around any of the remaining points in the set $\{0, 1, \infty, x_0, y_0\}$. It then travels back to x_0 or y_0. Five such loops are sufficient to generate $\pi_1(Q, (x_0, y_0))$ (see [145], p. 147).

1.3 COMPLEX MANIFOLDS AND METRICS

In this section, we go over some basics of the theory of complex manifolds and Hermitian metrics. We first recall the notion of real manifold. An m-dimensional smooth, connected *real manifold* X is a connected, topological Hausdorff space with a countable basis for its topology, together with a smooth (C^∞) structure. This smooth structure is given by an equivalence class of atlases. An atlas $\{U_i, h_i\}_{i \in I}$ is a family of open subsets U_i of X such that the U_i cover X, together with homeomorphisms $h_i : U_i \to V_i$ for each $i \in I$, the image V_i being an open subset of $\mathbb{R}^m$. We require that for every $i, j \in I$ such that $U_i \cap U_j \neq \phi$, the transition function $h_j \circ h_i^{-1}$ from $h_i(U_i \cap U_j)$ to $h_j(U_i \cap U_j)$ be a smooth diffeomorphism. Two atlases are compatible if their union is also an atlas, and this compatibility defines an equivalence relation, the equivalence classes being the smooth structures (see [14], §2.2). In even dimension $m = 2n$, the orientation on $\mathbb{R}^{2n}$, given by ordering the natural Euclidean coordinates by $x_1, y_1, \ldots, x_n, y_n$, defines an orientation on X provided the transition functions $h_j \circ h_i^{-1}$ are orientation preserving. Identifying $\mathbb{R}^{2n}$ with $\mathbb{C}^n$ using the complex coordinates $z^i = x_i + \sqrt{-1}y_i$ endows X with a complex structure provided the transition functions are holomorphic. Such transition functions have positive real Jacobian given by the square of the absolute value of the complex Jacobian, and hence are orientation preserving. Under these conditions, we say that X is a smooth, connected *complex manifold* of complex dimension n.

Let X be a complex manifold of dimension n. A *metric* on X is given locally by a tensor (sometimes written without the tensor product sign)

$$ds^2 = \sum_{k,\ell=1}^n g_{k\bar{\ell}} dz^k d\bar{z}^\ell = \sum_{k,\ell=1}^n g_{k\bar{\ell}} dz^k \otimes d\bar{z}^\ell,$$

where the $g_{k\bar{\ell}}$ are smooth functions. It is therefore a form g on the complex tangent space $T_z(X)$ to X at $z \in X$, with

$$g(\lambda \partial/\partial z^j, \mu \partial/\partial z^k)\,|_z := \lambda \overline{\mu} g_{j\bar{k}}(z),$$

where $z^1, \ldots, z^n$ are local coordinates at z and $g_{j\bar{k}}$ is smooth at z.

Assume the metric is *Hermitian*, that is, at every point the $n \times n$ matrix with (k, ℓ)-entry $g_{k\bar{\ell}}$ is positive definite and Hermitian ($g_{k\bar{\ell}} = \overline{g_{\ell\bar{k}}}$). Then the manifold X is called a *Hermitian manifold*. To such a metric we associate its $(1, 1)$-form,

$$\omega = \sqrt{-1} \sum_{k,\ell=1}^n g_{k\bar{\ell}} dz^k \wedge d\bar{z}^\ell.$$

The *fundamental form* of the metric is often defined to be the above form divided by 2π. As $g_{k\bar{\ell}} = \overline{g_{\ell\bar{k}}}$, we have $\omega = \overline{\omega}$, so that ω is a $(1, 1)$-form that is a real 2-form on X. Such a form is called a real $(1, 1)$-form. The metric is defined to be Kähler when the form ω is closed, that is, $d\omega = 0$. The form $\frac{\omega}{2\pi}$ is then called the Kähler form and is a representative of a class in $H^2(X, \mathbb{R})$. The concept of a Kähler metric was introduced by Eric Kähler in [73], where he also showed that such a metric can be written locally as

$$g_{k\bar{\ell}} = \partial^2 K / \partial z^k \partial \bar{z}^{\ell}$$

for some smooth real-valued function K, called the *Kähler potential*. It is easy to check directly that, conversely, any metric locally expressible in this form is Kähler.

We now give the two examples of Kähler metrics that we will encounter the most often in this book:

- Projective space $\mathbb{P}_N$ carries a metric, called the Fubini-Study metric, which is Kähler. The space $\mathbb{P}_N$ is given by the set of points in $\mathbb{C}^{N+1}$ whose coordinates are not all 0, modulo the equivalence relation

$$(z_0, z_1, \ldots, z_N) \simeq (\lambda z_0, \lambda z_1, \ldots, \lambda z_N), \qquad \lambda \neq 0, \quad \lambda \in \mathbb{C}.$$

  For homogeneous coordinates $[z_0 : z_1 : \ldots : z_N]$ of $\mathbb{P}_N$, representing the above equivalence class, the Fubini-Study metric is given in the coordinate patch $z_i \neq 0$ by the Hermitian matrix

$$FS_{k\bar{\ell}} = \frac{\left(1 + \sum_k |w_k|^2\right) \delta_{k\ell} - \overline{w}_k w_\ell}{\left(1 + \sum_k |w_k|^2\right)^2}$$

$$= \partial^2 \log\left(1 + \sum_k |w_k|^2\right) / \partial w_k \partial \overline{w}_\ell,$$

  where $w_k = z_k / z_i$, $k = 1, \ldots, N$, the corresponding affine coordinates.
- The complex two-dimensional ball, or complex 2-ball for short, is

$$B_2 := \{z = (z^1, z^2) : |z^1|^2 + |z^2|^2 < 1\}.$$

In the affine chart

$$U_0 := \{z = (z_0 : z_i : z_2) \in \mathbb{P}_2 : z_0 \neq 0\} \simeq \mathbb{C}^2$$

of the projective plane, with the (affine) origin $o := (1 : 0 : 0)$, the ball can be described in homogeneous coordinates by

$$B_2(U_0) := \{z \in U_0 : F(z, z) < 0\} \simeq B_2,$$

where F is the indefinite Hermitian form on $\mathbb{C}^3$ given by

$$F(z, w) := -z_0 \overline{w}_0 + z_1 \overline{w}_1 + z_2 \overline{w}_2 = {}^t \overline{w} S z \qquad (1.3)$$

(for $z = {}^t(z_0, z_1, z_2)$, $w = {}^t(w_0, w_1, w_2)$), and S is the associated matrix

$$S := \begin{pmatrix} -1 & 0 & 0 \\ 0 & 1 & 0 \\ 0 & 0 & 1 \end{pmatrix}.$$

The Hermitian form F on the ball $B_2(U_0)$, given in (1.3), defines a positive function (in homogeneous coordinates),

$$N(z) := -F(z, z)/z_0 \overline{z}_0.$$

In affine coordinates $z = (z^1, z^2) = (1 : z^1 : z^2)$, we have

$$N = N(z) = 1 - z^1 \overline{z}^1 - z^2 \overline{z}^2.$$

With the help of this "distance function" we define a Hermitian metric on the ball by

$$g = ds^2 = \sum g_{j\overline{k}} dz^j \, d\overline{z}^k$$

(as a form on the tangent space at z, we have $g(\lambda \partial/\partial z^j, \mu \partial/\partial z^k) \, |_z :=$ $\lambda \overline{\mu} g_{j\overline{k}}(z)$), where

$$g_{j\overline{k}} = g_{j\overline{k}}(z) := \frac{-\partial^2 \log N(z)}{\partial z^j \, \partial \overline{z}^k} = \left(N \delta_{jk} + \overline{z}^j z^k \right) / N^2. \qquad (1.4)$$

The *Kähler form* associated to this metric is

$$\omega = (i/2\pi) \sum g_{j\overline{l}} dz^k \wedge d\overline{z}^l = -(i/2\pi) \partial \overline{\partial} \log N,$$

which is closed and hence is a *Kähler form*. Therefore g defines a *Kähler metric*. This is a real form, in that $\omega = \overline{\omega}$, of type $(1, 1)$.

A real $(1, 1)$-form ω on a complex manifold Y is called *positive* if, for every holomorphic tangent vector v on Y, we have $-\sqrt{-1}\omega(v, \overline{v}) \geq 0$ and *positive definite* if, whenever $v \neq 0$, this inequality is strict. This enables us to define inequalities between real $(1, 1)$-forms. The fundamental form of a Hermitian metric is a positive $(1, 1)$-form since, on the holomorphic tangent space at $y \in Y$, we have, in local coordinates, $g(\lambda \partial/\partial z^k, \mu \partial/\partial z^\ell)_{|_y} = \lambda \overline{\mu} g_{k\overline{\ell}}(y)$.

On the other hand, we can associate to the metric its Ricci form, given locally by

$$\gamma_1 = \left(\sqrt{-1}/2\pi \right) \sum_{k,\ell} R_{k\overline{\ell}} dz^k \wedge d\overline{z}^\ell,$$

where

$$R_{k\overline{\ell}} = -\partial^2 \log \det \left(g_{k\overline{\ell}} \right) / \partial z^k \partial \overline{z}^\ell.$$

This is a closed real $(1, 1)$-form. It is also called the Chern form since, as we shall see in the next section, the first Chern class $[c_1(Y)] \in H^2(Y, \mathbb{Z})$ of Y is represented in $H^2(Y, \mathbb{R})$ by $[\gamma_1]$.

1.4 DIVISORS, LINE BUNDLES, THE FIRST CHERN CLASS

We collect here some general facts about divisors, line bundles, and the first Chern class on a complex manifold X. This material is used in Chapters 3 and 4, where, for the most part, X is a compact, smooth, connected algebraic surface, and in Chapter 6, where X is not necessarily compact and where we also incorporate an orbifold structure. The material is standard and can be found in many books on complex and algebraic manifolds. In particular, we do not define Čech cohomology; nor do we say much about de Rham cohomology beyond its definition, since we will ultimately be needing only very concrete special cases of these objects. The necessary background in singular cohomology can be found in §1.1. This foundational material is of course very important, but, for general remarks, we content ourselves with borrowing from the classical treatments in [49], Chapters 0 and 1, and [131], Chapter VI, referring the reader to these references for details of all proofs and justifications omitted here.

Let X be a complex manifold. We do not assume for the moment that X is compact. An analytic hypersurface Y on X is an analytic subvariety of codimension 1. Locally, it is given by the zeros of a single nonzero holomorphic function. An analytic hypersurface Y on X need not be smooth, and it is irreducible if and only if the locus of its non-singular points is connected. The group $\mathrm{Div}(X)$ of *divisors* D on X is the free abelian group of locally finite linear combinations $\sum_i m_i D_i$, $m_i \in \mathbb{Z}$, where D_i is an irreducible analytic hypersurface on X. The terminology "locally finite" means that, for any $x \in X$, there exists a neighborhood of x meeting only a finite number of the D_i

appearing in the divisor D. If X is compact, this means the sum is finite. If all the $m_i \geq 0$, the divisor is called *effective*, or nonnegative. We can describe divisors using open coverings of X. Let $\{U_\alpha, f_\alpha\}$ be an open covering $\{U_\alpha\}$ of X together with a nontrivial meromorphic function f_α on each U_α such that $f_\alpha f_\beta^{-1}$ is a nonzero holomorphic function on $U_\alpha \cap U_\beta$ and $\text{ord}_Y(f_\alpha) = \text{ord}_Y(f_\beta)$ for any irreducible analytic hypersurface Y on X. Here, we use the fact that we can unambiguously define the order $\text{ord}_Y(f_\alpha)$ of f_α on $U_\alpha \cap Y$. It equals $a - b$, where f_α has a zero of order a along Y and a pole of order b along Y. Then, we can form the divisor $\sum_Y \text{ord}_Y(f_\alpha)Y$, where for each Y we choose α with $Y \cap U_\alpha$ nonempty. Conversely, any divisor can be so obtained. Multiplication of the functions f_α corresponds to addition of divisors. If f is a meromorphic function on X, then the divisor of f, denoted by (f), is defined by $(f) = \sum_Y \text{ord}_Y(f)Y$. Given a holomorphic map $\iota : X \to V$ of complex manifolds, the pullback map $\iota^* : \text{Div}(V) \to \text{Div}(X)$ associates to every divisor given by $D = \{U_\alpha, f_\alpha\}$ on V the divisor $\iota^*(D) = \{\iota^{-1}(U_\alpha), \iota^*(f_\alpha)\}$ on X, which is well defined as long as $\iota(X)$ is not contained in D. We say that two divisors D and D' on X are linearly equivalent if $D - D' = (f)$ for a meromorphic function f on X. The divisor group modulo this equivalence is called the divisor class group. We denote the image of a divisor D in the divisor class group by $cl(D)$.

A divisor on X corresponds to a (holomorphic) line bundle on X. A line bundle $\pi : L \to X$ is given by $\{U_\alpha, \varphi_\alpha\}$, where $\{U_\alpha\}$ is an open cover of X and φ_α identifies $\pi^{-1}(U_\alpha)$ biholomorphically with $U_\alpha \times \mathbb{C}$. The transition functions $f_{\alpha\beta} : U_\alpha \cap U_\beta \to \mathbb{C}^*$ are given by $\varphi_\alpha \circ \varphi_\beta^{-1}$. It is immediate that the $f_{\alpha\beta}$ are holomorphic, nonvanishing, and satisfy $f_{\alpha\beta}f_{\beta\alpha} = 1$, $f_{\alpha\beta}f_{\beta\gamma}f_{\gamma\alpha} = 1$. If we are given $f_{\alpha\beta}$ on $U_\alpha \cap U_\beta$ with these properties, we can use them to patch together a line bundle, and another system of transitions functions $f'_{\alpha\beta}$ defines the same line bundle if and only if we have $f'_{\alpha\beta} = (f_\alpha)(f_\beta)^{-1}f_{\alpha\beta}$ for f_α nonzero holomorphic on U_α. Line bundles have a natural group structure with multiplication given by "tensor product" and inversion given by "duality." The tensor product corresponds to multiplication of transition functions and duality to their inversion. The group of line bundles on X is called the Picard group $\text{Pic}(X)$. Let $D = \{U_\alpha, f_\alpha\} \in \text{Div}(X)$. The $f_{\alpha\beta} = f_\alpha f_\beta^{-1}$ are holomorphic, nonvanishing, and satisfy $f_{\alpha\beta}f_{\beta\alpha} = 1$, and $f_{\alpha\beta}f_{\beta\gamma}f_{\gamma\alpha} = 1$, and therefore define a line bundle, independent of the local data for D, which we denote by $[D]$. The map $D \to [D]$ defines a group homomorphism from $\text{Div}(X)$ to $\text{Pic}(X)$, with kernel the divisors (f) of the meromorphic functions on X, and therefore a map $cl(D) \mapsto [D]$ from the divisor class group to $\text{Pic}(X)$. We denote $cl(D)$ also by $[D]$ from now on. If $\iota : X \to V$ is a holomorphic map to a complex manifold V, then $\iota^*([D]) = [\iota^*(D)]$, where a line bundle on V with transition functions $f_{\alpha\beta}$ on $U_\alpha \cap U_\beta$ pulls back to one on X with transition functions $\iota^*(f_{\alpha\beta})$ on $\iota^{-1}(U_\alpha) \cap \iota^{-1}(U_\beta)$. A (global) meromorphic section s of a line bundle $\pi : L \to X$ is, roughly speaking, a meromorphic function $s : X \mapsto L$ that satisfies $\pi \circ s = 1$. If L has transition functions $f_{\alpha\beta}$, the section s is given by meromorphic maps $s_\alpha : U_\alpha \to \mathbb{C}$, with $s_\alpha = f_{\alpha\beta} s_\beta$ on $U_\alpha \cap U_\beta$. If the s_α have no poles, then s is a holomorphic section. It can be easily shown that a line bundle L is $[D]$ for some divisor D on X if and only if it has a global meromorphic section and this global section is holomorphic if D is effective. A section s that is only defined on some, possibly proper, open set $U \subset X$ is called a local section.

By using the description of line bundles by transition functions, it can be shown that the group $\mathrm{Pic}(X)$ is isomorphic to the Čech cohomology group $H^1(X, \mathcal{O}^*)$, where $\mathcal{O}$ is the sheaf of holomorphic functions on X, and $\mathcal{O}^*$ is the subsheaf of nowhere vanishing holomorphic functions on X (for a definition of Čech cohomology, see [49], Chapter 0, §3.). The short exact sequence of sheaves, $0 \to \mathbb{Z} \to \mathcal{O} \to \mathcal{O}^* \to 0$, where $\mathcal{O} \to \mathcal{O}^*$ is given by the exponential map, then gives a boundary map in cohomology from $H^1(X, \mathcal{O}^*)$ to $H^2(X, \mathbb{Z})$, the singular cohomology defined in §1.1 of this chapter. The image $c_1(L)$ in $H^2(X, \mathbb{Z})$ of a line bundle $L \in H^1(X, \mathcal{O}^*)$ is called the first Chern class of L. Not all elements of $H^2(X, \mathbb{Z})$ are equal to $c_1(L)$ for some holomorphic line bundle L. The image of c_1 is called the Néron-Severi group $\mathrm{NS}(X)$. If $L = [D]$, we write $c_1(D)$ for $c_1([D])$. We have $c_1(L_1 \otimes L_2) = c_1(L_1) + c_1(L_2)$ and $c_1(L^*) = -c_1(L)$. Moreover, for a holomorphic map $\iota : X \to V$ between complex manifolds, we have $c_1(\iota^*(L)) = \iota^*(c_1(L))$. The de Rham cohomology space $H^p_{\mathrm{DR}}(X, \mathbb{R})$ is the quotient group of closed p-forms on a differential manifold X modulo exact p-forms [49], p. 23. Viewing $H^2(X, \mathbb{Z})$ inside $H^2(X, \mathbb{R})$ and using the de Rham theorem [49], p. 44, which implies that $H^2(X, \mathbb{R})$ and $H^2_{\mathrm{DR}}(X, \mathbb{R})$ are isomorphic, we see that $c_1(L)$ is represented in $H^2_{\mathrm{DR}}(X, \mathbb{R})$ by a differential form $\gamma_1(L)$, called the first Chern form. It can be shown that, over $\mathbb{C}$, the differential form $\gamma_1(L)$ is represented in $H^2_{\mathrm{DR}}(X) = H^2_{\mathrm{DR}}(X, \mathbb{C})$ by a real $(1, 1)$-form. A holomorphic line bundle is called positive if $c_1(L)$ is represented by a positive real $(1, 1)$-form in $H^2_{\mathrm{DR}}(X)$.

In terms of a Hermitian metric on a line bundle L, we have an explicit formula for this Chern form as follows. Let $L = \{U_\alpha, f_{\alpha\beta}\}$ be a line bundle. A Hermitian metric h on L is given by positive smooth functions h_α on each U_α, such that on $U_\alpha \cap U_\beta$ we have $h_\alpha = |f_{\beta\alpha}|^2 h_\beta$. On $L_{|U_\alpha} \simeq U_\alpha \times \mathbb{C}$, this means that, in each fiber $\{x\} \times \mathbb{C}$, we have the Hermitian inner product given by $\langle (x, \lambda), (x, \mu) \rangle_x = \lambda \overline{\mu} h_\alpha(x)$. Therefore, if s is a global holomorphic section of L, then $|s|^2 =: |s_\alpha|^2 h_\alpha$ is well defined. For any Hermitian metric h on L, define $\gamma_1(L, h) =: \frac{1}{2\pi i} \partial \overline{\partial} \log h_\alpha$ on U_α. Here, in terms of local coordinates z^i, we have $\partial = \sum_i (\partial/\partial z^i) dz^i$ and $\overline{\partial} = \sum_i (\partial/\partial \overline{z}^i) d\overline{z}^i$, so that $\gamma_1(L, h)$ is a real $(1, 1)$-form. To check that it is well defined, note that $\log h_\alpha - \log f_{\beta\alpha} - \log \overline{f}_{\beta\alpha} = \log h_\beta$ on $U_\alpha \cap U_\beta$, for some local branch of $\log f_{\beta\alpha}$. From $\overline{\partial} \log f_{\beta\alpha} = \partial \log \overline{f}_{\beta\alpha} = 0$, it follows that $\frac{1}{2\pi i} \partial \overline{\partial} \log h_\alpha = \frac{1}{2\pi i} \partial \overline{\partial} \log h_\beta$ on $U_\alpha \cap U_\beta$, as required. It can be shown that the class of $\gamma_1(L, h)$ in $H^2_{\mathrm{DR}}(X)$ is independent of the choice of metric h and represents $c_1(L)$. Therefore, a line bundle L is positive if and only if it admits a Hermitian metric h whose *Ricci form*, defined by $\mathrm{Ric}(h) = -\sqrt{-1}\partial\bar\partial \log h$, is positive definite. Then $\gamma_1(L, h) = \frac{1}{2\pi}\mathrm{Ric}(h)$ is the associated $(1, 1)$-form of a Kähler metric on X. We will discuss Kähler metrics further in Chapter 4. A complex manifold is said to be a *Hodge manifold* if it carries a positive line bundle.

In future chapters, we will in practice mostly use the first Chern form of one Hermitian line bundle, namely the canonical line bundle, and that in dimension 2, except when $X = \mathbb{P}_N$, when we need it for all N. In Chapter 6, we will be interested in so-called "log canonical" divisors and bundles. For a general complex manifold X of dimension n, the transition functions of the canonical line bundle are described as follows. Let U_α be an open set in X isomorphic to an open set in $\mathbb{C}^n$, and $z_\alpha^1, \ldots, z_\alpha^n$ local coordinates on U_α. If U_β is another such open set, let $f_{\alpha\beta}$ be the determinant of the $n \times n$ matrix with (i, j)-entry $\partial z_\beta^i / \partial z_\alpha^j$, $i, j = 1, \ldots, n$. Then the canonical line bundle is defined as that with associated transition function data $\{U_\alpha, f_{\alpha\beta}\}$. Suppose that X carries a meromorphic $(n, 0)$-form ω_n. Then, by its definition, ω_n is given in terms of the local coordinates on U_α by $f_\alpha dz_\alpha^1 \wedge \ldots \wedge dz_\alpha^n$, where f_α is meromorphic on U_α and $f_{\alpha\beta} = f_\alpha f_\beta^{-1} = \det(\partial z_\beta^i / \partial z_\alpha^j)$ on $U_\alpha \cap U_\beta$. Any two such meromorphic $(n, 0)$-forms differ by a meromorphic function. We call $\{U_\alpha, f_\alpha\}$ the canonical divisor. Its divisor class is independent of ω_n and is called the canonical class. The associated line bundle is called the canonical line bundle. When the context is clear, we use the notation K_X for the canonical divisor and its class, and also for the canonical line bundle. The dual line bundle K_X^* is called the anticanonical bundle. A smooth volume form Ω_v on X is a positive smooth (n, n)-form, given by a collection v of functions $v_\alpha > 0$ on each set U_α of an open covering of X, with

$$\Omega_v = v_\alpha \prod_{i=1}^{n} \sqrt{-1}\, dz_\alpha^i \wedge d\bar z_\alpha^i,$$

and $v_\alpha = |\det(\partial z_\beta^i / \partial z_\alpha^j)|^2 v_\beta$ on $U_\alpha \cap U_\beta$. Therefore, the v_α^{-1} give rise to a Hermitian metric on the canonical bundle K_X, or, equivalently, the v_α define a Hermitian metric on K_X^*. The *Ricci form* $\mathrm{Ric}\,\Omega_v$ of Ω_v is defined to be the real $(1, 1)$-form given locally by

$$\mathrm{Ric}\,\Omega_v = -\sqrt{-1}\,\partial\bar\partial \log \Omega_v$$

$$=: -\sqrt{-1} \sum_{k, \ell = 1, \ldots, n} \left(\partial_k \bar\partial_\ell \log v_\alpha \right) dz_\alpha^k \wedge d\bar z_\alpha^\ell,$$

where $\partial_k = \partial/\partial z_\alpha^k$, $\bar\partial_\ell = \partial/\partial \bar z_\alpha^\ell$. Therefore, denoting by v^{-1} the metric on K_X given by the v_α^{-1}, we have

$$\gamma_1(K_X, v^{-1}) = -\frac{1}{2\pi}\mathrm{Ric}(\Omega_v).$$

If there is a meromorphic $(n, 0)$-form ω_n on X, given locally by a collection f of functions f_α, we have $\Omega_{|f|^2} = \omega_n \wedge \overline{\omega}_n$. We define the first Chern class $c_1(X)$ of X to be $-c_1(K_X) = c_1(K_X^*)$. It is represented by the first Chern form $\gamma_1(X, v) =: -\gamma_1(K_X, v^{-1})$, for any choice of v.

When $X = \mathbb{P}_N$, with projective coordinates $[Z_0 : \ldots : Z_N]$, we can define a meromorphic $(n, 0)$-form, on the coordinate patch U_0 given by $Z_0 \neq 0$, as

$$\omega = w_1^{-1} dw_1 \wedge \ldots \wedge w_N^{-1} dw_N,$$

where $w_i = Z_i/Z_0$. The form ω is nonzero on this patch, with a single pole along each hyperplane $Z_i = 0$. On the coordinate patch U_j given by $Z_j \neq 0$, $j = 1, \ldots, N$, we have

$$\omega = (-1)^j w_0^{-1} dw_0 \wedge \ldots \wedge \widehat{w_j^{-1} dw_j} \wedge \ldots \wedge w_N^{-1} dw_N,$$

where $w_i = Z_i/Z_j$, and the notation $\widehat{w_j^{-1} dw_j}$ means that the term $w_j^{-1} dw_j$ is omitted. Therefore ω also has a single pole along $Z_0 = 0$. We have $\text{Pic}(\mathbb{P}_N)$ isomorphic to $H^2(\mathbb{P}_N, \mathbb{Z}) = \mathbb{Z}$, so that every divisor on $\mathbb{P}_N$ is linearly equivalent to a multiple of the hyperplane divisor $H = \mathbb{P}_{N-1} \subset \mathbb{P}_N$. The line bundle $[H]$ is called the hyperplane bundle. It follows that, for the canonical bundle, we have $K_{\mathbb{P}_N} = [-(N+1)H]$. Let H_0 denote the hyperplane $Z_0 = 0$ in $\mathbb{P}_N$. Then 1 is a local defining function for H_0 on U_0 and Z_0/Z_j are local defining functions for H_0 on U_j. The line bundle $[H_0] = [H]$ therefore has transition functions $f_{\alpha\beta} = Z_\beta/Z_\alpha$ on $U_\alpha \cap U_\beta$. It is often denoted by $\mathcal{O}_{\mathbb{P}_N}(1)$. Its dual is called the tautological bundle, and is denoted by $\mathcal{O}_{\mathbb{P}_N}(-1)$. On each U_j, we can set $g_j = |Z_j|^{-2}||Z||^2$, where $||Z||^2 = Z_0^2 + \ldots Z_N^2$. On $U_i \cap U_j$ we have $g_i = |Z_j/Z_i|^2 g_j$, so that the g_i^{-1} define a Hermitian metric g^{-1} on the line bundle $[H]$. Therefore $c_1(H)$ is represented by $\gamma_1(H, g^{-1}) = \frac{-1}{2\pi i} \partial\overline{\partial} \log ||Z||^2 = \frac{1}{2\pi} FS$, where $FS =: \sqrt{-1}\partial\overline{\partial} \log ||Z||^2$, which is a real closed positive $(1, 1)$-form on $\mathbb{P}_N$, and is the Fubini-Study Kähler form (up to 2π). It is the $(1, 1)$-form associated to the Fubini-Study Kähler metric on $\mathbb{P}_N$ (see §1.3 of this chapter). Therefore, the hyperplane bundle is positive. We see also that $c_1(K_{\mathbb{P}_N})$ is represented by the negative real $(1, 1)$ form $-(N+1)\frac{1}{2\pi} FS$.

We can describe the tautological line bundle in another way, which explains its name. The points of $\mathbb{P}_N$ are the equivalence classes of the elements of $\mathbb{C}^{N+1} \setminus \{0\}$ modulo scalar multiplication by the elements of $\mathbb{C}^*$. Therefore, each point $P \in \mathbb{P}_N$ corresponds to a copy of $\mathbb{C}^*$, since we can choose a representative $\widetilde{P} \neq 0$ of P in $\mathbb{C}^{N+1}$ and identify P with the points $\lambda\widetilde{P}, \lambda \in \mathbb{C}^*$. We then adjoin $\{0\}$ to $\mathbb{C}^*$ to get a copy of $\mathbb{C}$ at each point $P \in \mathbb{P}_N$. We then patch these copies of $\mathbb{C}$ together to form the tautological line bundle on $\mathbb{P}_N$.

Assume in this paragraph that X is a smooth compact connected algebraic complex manifold, and let $\pi : L \to X$ be a holomorphic line bundle on X. As X is compact, the vector space of (holomorphic) global sections of L is finite dimensional. A linear system on X is the vector space spanned by a set $\{s_0, \ldots, s_N\}$ of linearly independent global sections of L. When this vector space consists of all global sections of L, we call it a complete linear system $|L|$. Then, we have a rational map $\varphi_{|L|} : X \to \mathbb{P}_N$ given by $x \mapsto [s_0(x), \ldots, s_N(x)]$, where, $s_i(x) \in \mathbb{C}$ is the value of s_i at x in a local trivialization $L_{|U} \simeq U \times \mathbb{C}$. This map is defined on the complement in X of the common zero set of all the sections s_i and, up to an automorphism of $\mathbb{P}_N$, is independent of the choice of basis of the complete linear system. A line bundle L is said to be very ample if the rational map $\varphi_{|L|}$ determined by its complete linear system is an embedding. It is said to be ample if there is a positive integer m such that the line bundle $L^{\otimes m}$ is very ample. When L is very ample, we can view $s_0, \ldots, s_N$ as giving a coordinate system on X, embedded in $\mathbb{P}_N$ by $\varphi_{|L|}$. Indeed, define an open covering $\{V_\alpha\}$ of X by $V_\alpha = \{x \in X : s_\alpha(x) \neq 0\} = \varphi_{|L|}^{-1}(U_\alpha \cap \varphi_{|L|}(X))$ where U_α is the open set $Z_\alpha \neq 0$ in $\mathbb{P}_N$, and $[Z_0 : \ldots : Z_N]$ are the projective coordinates. The transition functions on $U_\alpha \cap U_\beta$ for the hyperplane bundle $[H]$ pull back to the transition functions on $V_\alpha \cap V_\beta$ for L. Therefore $L = \varphi_{|L|}^*[H] = \varphi_{|L|}^*(\mathcal{O}_{\mathbb{P}_N}(1))$. We can assume here that the V_α are the open sets of a trivialization of L.

By Kodaira's embedding theorem [49], p. 181, if L is a holomorphic positive line bundle on a complex compact manifold X, then it is ample, so that for all m sufficiently large ($m \gg 0$), the map $\varphi_{|L^{\otimes m}|}$ gives a holomorphic embedding of X into a projective space. Conversely, for $m \gg 0$, if L is ample there is a basis $\{s_0, \ldots, s_N\}$ of the global holomorphic sections of $L^{\otimes m}$ defining a map $\varphi_{|L^{\otimes m}|}$, as described above, that is a holomorphic embedding of X into the projective space $\mathbb{P}_N$. If $[H]$ is the hyperplane bundle of that projective space, then $L^{\otimes m}$ can be identified with the pullback $\varphi_{|L^{\otimes m}|}^*[H]$. Therefore, $c_1(L^{\otimes m}) = mc_1(L)$ is represented by $\frac{-1}{2\pi i} \partial \overline{\partial} \log ||h||^2$, where, for $x \in X$, we have

$$||h||^2(x) = \sum_i |s_i(x)|^2.$$

The $(1, 1)$-form $\sqrt{-1} \partial \overline{\partial} \log ||h||^2$ is positive and is the pullback of the Fubini-Study form on $\mathbb{P}_N$ defined above. Therefore the form representing $c_1(L)$ is positive, from which we see that a holomorphic line bundle L over a complex compact manifold is positive if and only if it is ample.

We shall be mainly interested in the situation of the last paragraph when $L = K_X$ or $K_X + D$, for some divisor D on X with rational coefficients. A divisor with rational coefficients is defined in the same way as a divisor earlier in this section, except that the coefficients of the irreducible divisors in its expression as a sum are allowed to be rational numbers, so such a divisor is in $\text{Div}(X) \otimes_\mathbb{Z} \mathbb{Q}$. Consider the case where $L = K_X$ is ample and let $m \gg 0$ be such that $K_X^{\otimes m}$ is very ample. Let φ_m be a corresponding embedding of X into a projective space $\mathbb{P}_N$. Then $c_1(K_X^{\otimes m})$ is represented by the form $\frac{1}{2\pi} \varphi_m^*(FS)$. The form $\varphi_m^*(FS)$ is the pullback by φ_m to X of the Fubini-Study form on $\mathbb{P}_N$, and is a positive real closed $(1, 1)$-form, as is $\omega = \frac{1}{m}\varphi_m^*(FS)$. From our previous discussion, for any volume form Ω on X, the class $c_1(K_X)$ is represented by $-\frac{1}{2\pi}\mathrm{Ric}(\Omega)$, so that there exists a smooth volume form Ω_v on X such that $\omega = -\mathrm{Ric}(\Omega_v)$.

In Chapters 5 and 6 we shall encounter the following situation. Consider a divisor D in $\mathrm{Div}(X) \otimes_{\mathbb{Z}} \mathbb{Q}$, where the support of D is an irreducible analytic subvariety of X (that is, for some integer s the divisor sD has just one term, an integer times an irreducible divisor). Suppose that $K_X + D$ is ample, meaning that there is an integer $m \gg 0$ such that $m(K_X + D)$ is a very ample divisor in $\mathrm{Div}(X)$. Let L be the line bundle $[m(K_X + D)]$ and $\varphi_m : X \to \mathbb{P}_N$ a projective embedding of X associated to L. Then $c_1(L) = c_1([m(K_X + D)])$ is represented by the closed real positive $(1, 1)$ form given by $\frac{1}{2\pi}\varphi_m^*(FS)$, where $\varphi_m^*(FS)$ is the pullback to X by φ_m of the Fubini-Study form on $\mathbb{P}_N$. Therefore, $\omega = \frac{1}{m}\varphi_m^*(FS)$ is a real closed positive $(1, 1)$-form such that $\frac{m}{2\pi}\omega$ represents $c_1([m(K_X + D))$. The Chern class $c_1([mK_X])$ is represented by $(-1/2\pi)m\mathrm{Ric}(\Omega)$ for a smooth volume form on X. We have $L = [mK_X][mD]$, and $c_1(mD)$ is represented by $\frac{1}{2\pi i}\partial\bar\partial \log h$ for any Hermitian metric h on $mD \in \mathrm{Div}(X)$. For a Hermitian metric h on mD and any holomorphic section $\sigma \neq 0$ of mD, the quantity $||\sigma||^2 =: |\sigma_\alpha(x)|^2 h_\alpha(x)$ is well defined. Moreover, outside the divisor of σ, we have $\partial\bar\partial \log ||\sigma||^2 = \partial\bar\partial \log h$. Therefore, outside the divisor of σ, the Chern class $c_1(mD)$ is represented by $-\sqrt{-1}\frac{1}{2\pi}\partial\bar\partial \log ||\sigma||^2$. There is therefore a smooth volume form Ω on X, a holomorphic section σ, and a Hermitian metric $|| \ ||$ on mD such that

$$\omega = -\mathrm{Ric}(\Phi) = \sqrt{-1}\partial\bar\partial \log(\Phi),$$

where Φ is the singular volume form

$$\Phi = \frac{\Omega}{||\sigma||^{2/m}},$$

wherever the right hand side exists.

This concludes the introductory material needed for later chapters. At times, some of the information of this chapter will be repeated so as to make our accounts more self-contained.

Chapter Two

Riemann Surfaces, Coverings, and Hypergeometric Functions

The theory of Riemann surfaces uses methods from many different branches of mathematics, including algebra, topology, geometry, and analysis. As we shall see, this is also true of the theory of complex surfaces. Moreover, many of the phenomena we consider for complex surfaces have their origin in the older study of Riemann surfaces. In this chapter, we review some basic concepts used in later chapters and, by way of example, apply them to Riemann surfaces. For the most part, we do not provide proofs, but we do provide individual references for the main facts. There are many good general references on Riemann surfaces and complex function theory, for example, [31], [39], [118].

We also present in this chapter some basic facts about the classical Gauss hypergeometric functions of one complex variable, whose generalization to Appell hypergeometric functions of two complex variables is the subject of Chapter 7. Again, our treatment is more an introduction to Chapter 7 than a thorough treatment of the one-variable case. For more details, there are many references, for example, [25], [145].

A word of warning about terminology: the word "surface" in this chapter refers to real dimension 2, whereas in later chapters the word "surface" will refer to complex dimension 2, that is, real dimension 4. When we consider a complex one-dimensional algebraic variety, it is referred to as a "curve," but at the same time it has an underlying real two-dimensional structure. In this way the words "curve" and "surface" are associated to the same object due to the passage from the complex numbers to the reals.

2.1 GENUS AND EULER NUMBER

A *Riemann surface* is a one-dimensional connected complex manifold. It thus has the underlying structure of a real two-dimensional connected oriented smooth manifold (for definitions, see Chapter 1). For a Riemann surface we need not require that the topology have a countable basis: this follows from a theorem of Radó, which also states that a Riemann surface can be triangulated (see the essay of Remmert and Schneider in [142], p. 188).

Recall from Chapter 1 that a path-connected space is simply connected if any simple closed curve can be continuously deformed to a point within the space. An important result, known as the *Uniformization Theorem*, proved independently by Koebe and Poincaré in 1907 (see [39], p. 194), describes the simply-connected Riemann surfaces. It states that, up to isomorphism, there are three simply-connected Riemann surfaces. They are the complex numbers $\mathbb{C}$, the complex projective line $\mathbb{P}_1$, and the upper half plane $\mathcal{H}$. Recall that the points of $\mathbb{P}_1$ are given by homogeneous complex coordinates $(z_1 : z_2) \neq (0 : 0)$ with the equivalence relation $(z_1 : z_2) = (\lambda z_1 : \lambda z_2)$, for all nonzero complex numbers λ. Therefore, $\mathbb{P}_1$ is the same as $\mathbb{C}$ together with a one-point compactification at $\infty = (1 : 0)$. The projective line is often referred to as the *Riemann sphere*, as it can be represented as the sphere S^2 of radius 1 in $\mathbb{R}^3$. By stereographic projection from the north pole of S^2 onto the tangent plane at the south pole, identified with the complex line, we recover $\mathbb{P}_1$. The north pole goes to the point ∞ on $\mathbb{P}_1$. Notice that of the three inequivalent simply-connected Riemann surfaces only $\mathbb{P}_1$ is compact.

The upper half plane $\mathcal{H}$ comprises the complex numbers with positive imaginary part. There is a biholomorphic map from $\mathcal{H}$ to the unit disk $D : |w| < 1$, given by

$$z \mapsto w = \frac{z - i}{z + i}.$$

A simply-connected Riemann surface X has a trivial fundamental group $\pi_1(X)$. This is the group of all homotopy classes of closed paths that begin and end at a fixed base point with multiplication given by path composition (see Chapter 1, §1.2). To abelianize $\pi_1(X)$, we must factor out by the commutator subgroup $[\pi_1(X), \pi_1(X)]$, thus obtaining the first homology group,

$$H_1(X, \mathbb{Z}) = \pi_1(X)/[\pi_1(X), \pi_1(X)].$$

Recall from Chapter 1, §1.1, that the real manifold underlying a compact Riemann surface is homeomorphic to S^2 with g handles attached, and that the invariant g is called the *genus* of the Riemann surface. It gives a complete classification of orientable compact smooth surfaces. The Euler number of a Riemann surface F_g of genus g equals $2 - 2g$. Notice that

$$e(F_g \setminus \{p_1, \ldots, p_r\}) = 2 - 2g - r,$$

where the p_i are r points of F_g. (Note that the space $F_g \setminus \{p_1, \ldots, p_r\}$ is homotopically equivalent to F_g with r compact disks removed; then use Mayer-Vietoris from Chapter 1, §1.1.) This gives a negative Euler number when $g = 0$ and $r \geq 3$, when $g = 1$ and $r \geq 1$, and when $g \geq 2$ and r is arbitrary.

The raison d'être of Riemann surfaces are the branched coverings of $\mathbb{P}_1$ on which multivalued functions can be made univalued. In later chapters, complex surfaces, bearing the same relation to $\mathbb{P}_2$, will be the main players in our story. For now, consider the example of four distinct points $a_i, i = 1, \ldots, 4$, on $\mathbb{P}_1$, and the multivalued function

$$\sqrt{(z - a_1)(z - a_2)(z - a_3)(z - a_4)}.$$

This defines a two-sheeted covering of $\mathbb{P}_1$ with four branch points at the a_i where two sheets come together and where $t = \sqrt{z - a_i}$ is a local coordinate. Another example is given by the multivalued function

$$\sqrt[6]{(z - a)^2(z - b)^3}, \qquad a \neq b.$$

This defines a branched covering of $\mathbb{P}_1$ of degree 6. Over a we have two points on the Riemann surface where three sheets come together, over b we have three points where two sheets come together, and over ∞ we have one point where six sheets come together. Over each of the remaining points of $\mathbb{P}_1$ we have six points on the Riemann surface. In both cases, the genus of the Riemann surface is 1.

2.2 MÖBIUS TRANSFORMATIONS

Our first examples of Riemann surfaces were the simply-connected Riemann surfaces, which are, up to isomorphism, $\mathbb{P}_1$, $\mathbb{C}$, or $\mathcal{H}$. The Riemann sphere S^2 and $\mathbb{R}^2$ admit exactly one complex structure. The torus, F_1, has infinitely many complex structures parameterized by $\mathrm{PSL}_2(\mathbb{Z})\backslash\mathcal{H}$, and F_g ($g \geq 2$) has infinitely many complex structures parameterized by a space of complex dimension $3g - 3$ (see [133], p. 117).

An automorphism of a Riemann surface is a biholomorphic map of the Riemann surface onto itself. Both $\mathbb{C}$ and $\mathcal{H}$ are subspaces of $\mathbb{P}_1$, and, in fact, the automorphism groups of these simply-connected Riemann surfaces are all subgroups of

$$\mathrm{Aut}\,(\mathbb{P}_1) = \mathrm{PGL}(2, \mathbb{C}) = \mathrm{GL}(2, \mathbb{C})/\mathbb{C}^* \qquad (2.1)$$

acting by fractional linear transformations. That is, using projective coordinates $(z_1 : z_2)$, we have

$$(z_1 : z_2) \mapsto (az_1 + bz_2 : cz_1 + dz_2), \qquad \begin{pmatrix} a & b \\ c & d \end{pmatrix} \in \mathrm{GL}(2, \mathbb{C}).$$

On multiplying the associated 2×2 matrix by a scalar factor if necessary, we may assume these automorphisms are in PSL(2, $\mathbb{C}$), where

$$\mathrm{PSL}(2, \mathbb{C}) = \mathrm{SL}(2, \mathbb{C})/\{\pm 1\}. \tag{2.2}$$

We have

$$\mathrm{Aut}\,(\mathbb{C}) = \{z \mapsto az + b; a \neq 0, b \in \mathbb{C}\}. \tag{2.3}$$

The group

$$\mathrm{Aut}\,(\mathcal{H}) = \mathrm{PSL}(2, \mathbb{R}) = \mathrm{SL}(2, \mathbb{R})/\{\pm 1\} \tag{2.4}$$

acts by fractional linear transformations on $\mathcal{H}$. That is, for $z \in \mathcal{H}$,

$$z \mapsto \frac{az + b}{cz + d}, \qquad \begin{pmatrix} a & b \\ c & d \end{pmatrix} \in \mathrm{SL}(2, \mathbb{R}). \tag{2.5}$$

There is a biholomorphic map from $\mathcal{H}$ to the unit disk D given by

$$z \mapsto \frac{z - i}{z + i}.$$

Using projective coordinates, we have $(z_1 : z_2) \in D$ whenever

$$|z_1|^2 - |z_2|^2 < 0,$$

and this inequality defines a Hermitian form of type $(1, -1)$, which is invariant under the unitary subgroup $U(1, 1)$ of GL(2, $\mathbb{C}$). Moreover,

$$\mathrm{Aut}\,(D) = \mathrm{PU}(1, 1) \subset \mathrm{PGL}(2, \mathbb{C}). \tag{2.6}$$

Depending on the situation, we choose for the most convenient equivalent description either $\mathcal{H}$ or D.

Proofs of the above facts on automorphism groups are given in [118]. To summarize, an automorphism of $\mathbb{C}$ cannot have an essential singularity at ∞ by the theorem of Casorati-Weierstrass ([118], p. 242). Therefore, it is a polynomial and, by bijectivity, it must be linear. An automorphism of $\mathbb{P}_1$ can be transformed by fractional linear transformations to an automorphism fixing ∞. This must be linear. Therefore, the original automorphism is in PSL(2, $\mathbb{C}$). An automorphism of the unit disk can be transformed, by composing with elements of PU(1, 1), into an automorphism of the disk fixing the origin ([118], p. 71). Such an automorphism is a rotation, by Schwarz's Lemma ([118], p. 212).

Let

$$\gamma = \begin{pmatrix} a & b \\ c & d \end{pmatrix}$$

be a typical element of SL(2, $\mathbb{C}$). An invariant of the conjugacy classes in SL(2, $\mathbb{C}$) is given by the trace

$$\mathrm{tr}(\gamma) = a + d.$$

In PSL(2, $\mathbb{C}$) an invariant of the conjugacy classes is given by the square of the trace

$$\mathrm{tr}^2(\gamma) = (a + d)^2.$$

The fixed points of γ are the solutions z of the quadratic equation

$$\frac{az + b}{cz + d} = z,$$

which may be rewritten

$$cz^2 + (d - a)z - b = 0.$$

If $c = 0$ and $a = d = \pm 1$, then either $b = 0$ and we have the identity Möbius transformation or $b \neq 0$ and the transformation has one fixed point at infinity. In both cases, of course, $(a + d)^2 = 4$. If $c = 0$ and $a = 1/d \neq \pm 1$, the transformation is $z \mapsto a^2 z + ab$, which has two fixed points (one finite and one at ∞), since $a^2 \neq 1$.

If $c \neq 0$, then the solutions are

$$z = \frac{(a - d) \pm \sqrt{(a + d)^2 - 4}}{2c}.$$

There is one solution if $(a + d)^2 = 4$ and there are two solutions otherwise.

Therefore, every element of PSL(2, $\mathbb{C}$) that is not in the class of the identity has either one or two fixed points, according to whether $(a + d)^2 = 4$ or not.

These remarks lead us to the classification of Möbius transformations into types, which may be done using the square of the trace. A nontrivial Möbius transformation γ satisfying $\mathrm{tr}^2(\gamma) \in \mathbb{R}$ is called

(i) elliptic if $\mathrm{tr}^2(\gamma) < 4$,
(ii) parabolic if $\mathrm{tr}^2(\gamma) = 4$,
(iii) hyperbolic if $\mathrm{tr}^2(\gamma) > 4$.

The non-elliptic nontrivial transformations with two fixed points are called loxodromic. Hence, the non-hyperbolic loxodromic transformations have complex-valued square trace. It is straightforward to see that every parabolic transformation is conjugate in PGL$(2, \mathbb{C})$ to the transformation given by $z \mapsto z + 1$. The other transformations have two fixed points. After conjugating the transformation if necessary, we may send these fixed points to 0 and ∞, so that the transformation becomes

$$z \mapsto \rho^2 z, \qquad \rho \in \mathbb{C}.$$

There are two important cases to consider: nontrivial rotations with $\rho^2 = e^{i\theta}$, $\theta \in \mathbb{R} \setminus 2\pi\mathbb{Z}$, which has square trace $4\cos^2(\theta/2)$, and nontrivial dilatations with $\rho^2 = \lambda, \lambda > 0, \lambda \neq 1$, which has square trace $2 + \lambda + \lambda^{-1}$. Elliptic transformations are those conjugate to rotations. Hyperbolic transformations are those conjugate to dilatations. If γ is a nontrivial Möbius transformation with finite order n, then it is conjugate to the rotation

$$z \mapsto e^{2\pi i p/n} z, \qquad p \in (\mathbb{Z}/n\mathbb{Z})^*.$$

Except for this case, the transformations γ and γ^n have the same number of fixed points and are of the same type.

A transformation γ of PSL$(2, \mathbb{C})$ fixes a disk if and only if $\operatorname{tr}^2(\gamma) \geq 0$. For example, the transformation $z \mapsto z + 1$ maps the upper half plane to itself; the transformation $z \mapsto e^{i\theta} z, \theta \in \mathbb{R}$, leaves fixed the unit disk; and the transformation $z \mapsto \lambda z, \lambda > 0, \lambda \neq 1$, leaves invariant the half planes with boundary any line passing through the origin. A loxodromic non-hyperbolic transformation has no fixed disk. We shall not be concerned with such transformations. If $\gamma \neq \operatorname{Id}$ is a Möbius transformation with $\operatorname{tr}^2(\gamma) \geq 0$ and if z is not a fixed point of γ, then there is a circle passing through z such that the circle and the two disks bounded by it are invariant by γ. For proofs of all these statements, see [89] and [98].

Every Riemann surface is the quotient of one of the simply-connected Riemann surfaces $\mathcal{D}$ by a subgroup Γ of Aut$(\mathcal{D})$ acting freely discontinuously on $\mathcal{D}$. The action of a group Γ of homeomorphisms of a topological space $\mathcal{D}$ is freely discontinuous at a point $x \in \mathcal{D}$ if there exists a neighborhood U of x such that $g(U) \cap U = \phi$ for all $g \in \Gamma, g \neq \operatorname{Id}$. The action of Γ is freely discontinuous on $\mathcal{D}$ if it is freely discontinuous at each point of $\mathcal{D}$. The weaker condition that Γ act discontinuously at a point $x \in \mathcal{D}$ means that the set

$$\Gamma_x = \{g \in \Gamma \mid g(x) = x\}$$

is finite and that there is a neighborhood U of x such that $g(U) = U$ for all $g \in \Gamma_x$ and $g(U) \cap U = \phi$ for all $g \in \Gamma \setminus \Gamma_x$. The group acts discontinuously on $\mathcal{D}$ if it acts discontinuously at all points of $\mathcal{D}$.

A topological group is a set G that is both a group and a topological space on which the map

$$G \times G \to G,$$

$$(g, h) \mapsto gh^{-1}$$

is continuous. The group $SL(2, \mathbb{C})$ is a topological group with topology induced by the natural embedding of $SL(2, \mathbb{C})$ as a closed subspace of $\mathbb{C}^4$. This induces the quotient topology on $PSL(2, \mathbb{C})$. A subgroup of $PSL(2, \mathbb{C})$ is called discrete if it is a discrete subset of the topological group $PSL(2, \mathbb{C})$. Every finite subgroup of $PSL(2, \mathbb{C})$ is discrete. Every subgroup of $PSL(2, \mathbb{C})$ acting discontinuously on some nonempty set is discrete, but the converse does not hold. Still, a discrete subgroup of $PSL(2, \mathbb{R})$ acts discontinuously on all of $\mathcal{H}$ (see [89], Theorem 1.1).

2.3 METRIC AND CURVATURE

A (smooth) Riemannian metric on a (smooth real) manifold is a symmetric, bilinear, positive definite form in each tangent space that varies smoothly from point to point. Each of the simply-connected Riemann surfaces carries a natural Riemannian metric. On $\mathbb{P}_1$, viewed as the unit sphere S^2 in $\mathbb{R}^3$, we have the metric induced by the usual Euclidean metric on $\mathbb{R}^3$, which is invariant under the action of $SO(3)$. Moreover,

$$PU(2) = SU(2)/\{\pm 1\} = S^3/\{\pm 1\} = SO(3).$$

In terms of a complex coordinate $z = x + iy$ on $\mathbb{C}$ (via the stereographic projection), this metric on $\mathbb{P}_1 \setminus \{\infty\}$ is given by

$$ds^2 = \frac{4(dx^2 + dy^2)}{(1 + |z|^2)^2}. \tag{2.7}$$

At infinity, invariance leads to an expression in terms of $t = \frac{1}{z}$. On $\mathbb{C}$ we have the usual Euclidean metric

$$ds^2 = dx^2 + dy^2 \tag{2.8}$$

which is invariant under automorphisms of the form

$$z \mapsto az + b, \qquad |a| = 1.$$

Finally, on $\mathcal{H}$ we have the metric given in terms of the coordinate $z = x + iy$ by

$$ds^2 = \frac{dx^2 + dy^2}{y^2}, \tag{2.9}$$

which is invariant under the full automorphism group of $\mathcal{H}$. Alternatively, on the unit disk $\{z \in \mathbb{C} : |z| < 1\}$, the metric is given by

$$ds^2 = \frac{4dz \otimes d\bar{z}}{(1 - |z|^2)^2}. \tag{2.10}$$

Let X be an arbitrary Riemann surface. Then the universal covering space $\widetilde{X}$ of X is one of the three simply-connected Riemann surfaces and the group of covering transformations are isometries of the metrics introduced above. We can therefore project these metrics onto a metric on X, which we denote in terms of the local coordinate $z = x + iy$ by

$$ds^2 = g(z, \bar{z})dz \otimes d\bar{z} = g(z, \bar{z})(dx^2 + dy^2). \tag{2.11}$$

The corresponding surface element is

$$dF = g(z, \bar{z})dxdy \tag{2.12}$$

and the (Gaussian) curvature κ at every point is given by

$$\kappa = -\frac{2\partial_z \partial_{\bar{z}} \log g}{g}, \tag{2.13}$$

where $\partial_z = \frac{\partial}{\partial z}$ and $\partial_{\bar{z}} = \frac{\partial}{\partial \bar{z}}$. We have

$$\kappa dF = \frac{1}{i}\partial\bar{\partial} \log g, \tag{2.14}$$

where $\partial f = (\partial_z f)dz$ and $\bar{\partial} f = (\partial_{\bar{z}} f)d\bar{z}$. Direct computation in $\widetilde{X}$ shows that the curvature of the induced metric on X is constant and positive for $\widetilde{X} = \mathbb{P}_1$ (spherical), zero for $\widetilde{X} = \mathbb{C}$ (Euclidean), and negative for $\widetilde{X} = \mathcal{H}$ (hyperbolic) (see [39], pp. 213–220). The terms Euclidean, spherical, and hyperbolic refer to the respective geometries. For triangles in X whose sides are geodesics with respect to the metric ds^2, the angle sum is π in the Euclidean case, exceeds π in the spherical case, and is less than π in the hyperbolic case. The triangle with angles $\frac{\pi}{2}$, $\frac{\pi}{2}$, 0 occurs in the Euclidean case, and the vertex with 0 angle is at the point at infinity in the extended complex plane. In the hyperbolic case, triangles with 0 angles also occur, and the vertices whose angles are 0 are on the real line or at the point at infinity. By one formulation of the Gauss-Bonnet theorem, if Δ is a triangle with angles α, β, γ, in all cases we have

$$\alpha + \beta + \gamma - \pi = \int_\Delta \kappa\, dF.$$

In the non-Euclidean case the absolute value of this quantity is the area of the triangle (for more details about these remarks, see [39], p. 218, and [89], p. 65).

For example, for a triangle on S^2 with a vertex of angle α at the North Pole and two vertices meeting the equator at right angles, we have

$$\int_\Delta \kappa\, dF = \alpha.$$

As another example, consider the modular group $\text{PSL}_2(\mathbb{Z})$ generated by the fractional linear transformations given by the matrices

$$\begin{pmatrix} 0 & -1 \\ 1 & 0 \end{pmatrix} \qquad\qquad \begin{pmatrix} 1 & 1 \\ 0 & 1 \end{pmatrix}$$

and acting on $\mathcal{H}$. It has a fundamental region whose closure consists of two neighboring copies of a triangle T with angles $\frac{\pi}{2}, \frac{\pi}{3}, 0$. This fundamental region therefore has area $\frac{\pi}{3}$. Factoring $\mathcal{H}$ by the action of the modular group and compactifying by adding a point, we obtain a Riemann surface topologically equivalent to S^2, which we could think of as inheriting the volume $\frac{\pi}{3}$. However, S^2 is also topologically equivalent to the point compactification of the quotient of $\mathcal{H}$ by other so-called (hyperbolic) triangle groups of which $\text{PSL}_2(\mathbb{Z})$ is a special case. The closure of their fundamental domains consists of two neighboring copies of a hyperbolic triangle with angles $\frac{\pi}{p}, \frac{\pi}{q}, \frac{\pi}{r}$ where p, q, r are positive (possibly infinite) integers with $\frac{1}{p} + \frac{1}{q} + \frac{1}{r} < 1$. The area of this fundamental domain is then

$$2\pi \left| \frac{1}{p} + \frac{1}{q} + \frac{1}{r} - 1 \right|,$$

so S^2 inherits infinitely many different volumes by forming quotients of $\mathcal{H}$ by triangle groups. We shall encounter these triangle groups again later in this chapter.

The Gauss-Bonnet theorem can be seen as a relation between the Gaussian curvature of a compact Riemann surface X and its Euler characteristic. Let S be a convex polygon in X with r vertices and geodesic sides. Let α_v be the angle at the vertex v of S, and $\beta_v = \pi - \alpha_v$. Then, by one formulation of the Gauss-Bonnet theorem,

$$\int_S \kappa\, dF + \sum_v \beta_v = 2\pi, \tag{2.15}$$

which we can rewrite as

$$\sum_{v} \alpha_v - (r - 2)\pi = \int_S \kappa \, dF. \tag{2.16}$$

The angle sum of a Euclidean r-gon is $(r - 2)\pi$, so the curvature measures the difference of the angle sum from the Euclidean one. Now consider a cellular decomposition of X with all sides geodesics. Let e be the number of vertices, k the number of edges, and f the number of faces of the cellular decomposition. Let f_r be the number of r-gons. Then

$$\sum_{r} r f_r = 2k,$$

so, summing over all cells, we have

$$2\pi (e - k + f) = \int_X \kappa \, dF.$$

Recalling that this gives the Euler number, we have another formulation of the Gauss-Bonnet theorem:

$$e(X) = e - k + f = \frac{1}{2\pi} \int_X \kappa \, dF = \frac{1}{2\pi \sqrt{-1}} \int_X \partial \bar{\partial} \log g. \tag{2.17}$$

By using the geodesic curvature, we do not need to assume that the sides of the cellular decomposition are geodesics. The above proof can still be carried out as the terms with opposite signs coming from the sides cancel. It is convenient to think of $e(X)$ as measuring the volume of X.

Definition 2.1 *Let X be a compact Riemann surface; then the Euler volume of X is given by*

$$\text{vol}(X) = e(X) = \frac{1}{2\pi} \int_X \kappa \, dF,$$

where dF is the surface element of X.

Observe that the Euler volume may be negative. For example, an r-gon S in X with angles $\frac{\pi}{n_1}, \ldots, \frac{\pi}{n_r}$ and geodesic sides has Euler volume

$$\text{vol}(S) = \frac{1}{2} \left(\frac{1}{n_1} + \frac{1}{n_2} + \ldots + \frac{1}{n_r} - (r - 2) \right). \tag{2.18}$$

This formula remains valid for a polygon with geodesic sides in one of the simply-connected domains, with possibly some 0 angles at vertices on their boundary.

2.4 BEHAVIOR OF THE EULER NUMBER UNDER FINITE COVERING

In our discussions of complex surfaces and their coverings in later chapters, the behavior of topological invariants of these surfaces under finite coverings will play a crucial role. For Riemann surfaces we can study the behavior of the Euler number under a finite cover, which leads to the Hurwitz Formula. Let X and Y be compact connected Riemann surfaces with Euler numbers $e(X)$ and $e(Y)$, and let

$$f : Y \to X$$

be a finite cover, so f is a holomorphic surjective map such that each point of X has a pre-image of finite cardinality. Then at each point P of Y there is a local complex coordinate z such that $w = z^r$ is a local coordinate at $f(P)$ for some positive integer r. The map f is said to have ramification index $v_P(Y) = r - 1$ over $f(P)$, and the total ramification index of f is given by the finite sum

$$w = \sum_{P \in Y} v_P(Y). \tag{2.19}$$

We define the degree N of f to be the number of elements in the inverse image under f of a general (i.e., not a ramification) point. Therefore, for any $Q \in X$,

$$N = \sum_{P \in f^{-1}(Q)} (v_P(Y) + 1). \tag{2.20}$$

As we have seen, the Euler number behaves as the cardinality of a set. In a disk on X where f has no ramification, the intersection with any cellular decomposition will lift to N copies of itself on Y. However, compared to the unramified case, in a disk neighborhood of a point where f has ramification r, $r - 1$ points are lost in the lift. We must therefore subtract $r - 1$ from the Euler number, since

$$e(A \cup B) = e(A) + e(B) - e(A \cap B). \tag{2.21}$$

From this we derive the *Hurwitz formula*,

$$e(Y) = Ne(X) - w. \tag{2.22}$$

Consider the following example. Let

$$f : Y \to \mathbb{P}_1$$

be a covering of the projective line of degree N ramified over r distinct points $x_1, \ldots, x_r$. Assume that for $i = 1, \ldots, r$ there are $\frac{N}{n_i}$ points in $f^{-1}(x_i)$ each with multiplicity $n_i > 1$. Then

$$w = \sum_{i=1}^{r} \frac{N}{n_i}(n_i - 1), \tag{2.23}$$

and hence, by (2.22),

$$e(Y)/N = -\left(r - 2 - \sum_{i=1}^{r} \frac{1}{n_i}\right). \tag{2.24}$$

Notice the resemblance to the formula for the Euler volume of an r-gon with angles $\frac{\pi}{n_i}, i = 1, \ldots, r$. We have $Y = \mathbb{P}_1$ precisely when $e(Y) > 0$, that is, when

$$\sum_{i=1}^{r} \frac{1}{n_i} > r - 2. \tag{2.25}$$

Notice that, as $e(Y) = 2$, the degree N is determined by the n_i. For $r \geq 4$, it is not possible to satisfy the above inequality. Therefore, we must have $r \leq 3$, the interesting case being $r = 3$. For $r = 3$, we have the list of solutions $\frac{1}{2}, \frac{1}{2}, \frac{1}{k}, k \geq 2 ; \frac{1}{2}, \frac{1}{3}, \frac{1}{3} ; \frac{1}{2}, \frac{1}{3}, \frac{1}{4} ; \frac{1}{2}, \frac{1}{3}, \frac{1}{5}.$ We shall encounter the case $r = 3$ again in the next section of this chapter.

2.5 FINITE SUBGROUPS OF PSL(2, $\mathbb{C}$)

We have seen that every Riemann surface is the quotient of one of the simply-connected Riemann surfaces $\mathcal{D}$ by a subgroup Γ of Aut($\mathcal{D}$) acting freely discontinuously on $\mathcal{D}$. More generally, a finitely marked compact Riemann surface $\overline{S}$ is a compact Riemann surface together with a finite set of points on $\overline{S}$. To each such point $x \in \overline{S}$, one associates an integer $p_x, 2 \leq p_x \leq \infty$, called the branching number or order of x. Let S be the Riemann surface obtained from $\overline{S}$ by deleting those x with $p_x = \infty$. If the set of such $x \in \overline{S}$ is not empty, then S is not compact. The surface S has a simply-connected branched universal covering $\mathcal{D}$, where if $p_x < \infty$, then p_x is the branching number at x. The surface S is the quotient space $S = \mathcal{D}/\Gamma$, where Γ is a subgroup of Aut($\mathcal{D}$) acting discontinuously with only finite order fixed points in $\mathcal{D}$. For any point of $x \in S$, the branching number p_x equals the order of the stabilizer in Γ of a point lying over x. The basic signature of Γ is, by definition, $(g, n; p_1, \ldots, p_n)$, where g is the genus of $\overline{S}$ and n is the number of points of $\overline{S}$ with ramification at least 2, so $2 \leq p_i \leq \infty$. Of course, we can take $p_1 \leq \ldots \leq p_n$.

If $\mathcal{D} = \mathcal{H}$, then Γ is a discrete subgroup of $\mathrm{PSL}(2, \mathbb{R})$. If Γ has basic signature $(g, n; p_1, \ldots, p_n)$, then it has the presentation

$$\Gamma = \langle s_1, \ldots, s_n, u_1, v_1, \ldots, u_g, v_g : \prod_{j=1}^{n} s_j \prod_{m=1}^{g} [u_m, v_m] = s_1^{p_1} = \ldots = s_n^{p_n} = 1 \rangle,$$

where we omit the relation $s_j^{p_j} = 1$ if $p_j = \infty$ and where $[u, v] := uvu^{-1}v^{-1}$.

In this chapter, we shall mainly be concerned with discrete subgroups of $\mathrm{PSL}(2, \mathbb{R})$ with $g = 0$ and $n = 3$.

Pairs $(\mathcal{D}, \Gamma)$ with Γ finite appear in the work of Schwarz [128], mentioned above. He proved the following theorem (see [98] for a modern treatment).

Theorem 2.1 *Let Γ be a finite subgroup of the group of fractional linear transformations $\mathrm{PSL}(2, \mathbb{C})$. Then Γ has one of the following basic signatures:*

$(0, 2; p, p)$	cyclic
$(0, 3; 2, 2, p)$	dihedral
$(0, 3; 2, 3, 3)$	tetrahedral
$(0, 3; 2, 3, 4)$	cubic, octahedral
$(0, 3; 2, 3, 5)$	dodecahedral, icosahedral

where p is any integer with $2 \le p < \infty$.

The last three groups in the above list correspond to groups of symmetries of regular solids that preserve orientation. The octahedron is the dual of the cube, and the icosahedron the dual of the dodecahedron. In these cases, the group Γ is generated by a rotation R_1 in $\mathbb{R}^3$ fixing a face F of the solid, and a rotation R_2 in $\mathbb{R}^3$ fixing an edge E of F. The fixed point of R_2 is at the midpoint of E, the fixed point of R_1 is at the center of F, and $R_1 \circ R_2$ fixes an endpoint of E. A fundamental domain for the action induced on $S^2 \subset \mathbb{R}^3$ with the regular solid inscribed is given by the radial projection from the origin of S^2 of a triangle with one vertex at the center of F and the other two at the endpoints of E. This fundamental domain is then two adjacent copies of a spherical triangle with angles $\frac{\pi}{p}, \frac{\pi}{q}, \frac{\pi}{r}$ whose sum is greater than π: here, $(0, 3; p, q, r)$ is the signature of Γ. The cyclic group of order p given by the first group in the list is given by the transformation $z \mapsto e^{2\pi i/p} z$ and has fixed points 0 and ∞. The dihedral group of order $2p$ acts on $\mathbb{P}_1$ and is generated by the transformations $z \mapsto \frac{1}{z}$ and $z \mapsto e^{2\pi i/p} z$. A fundamental domain on S^2 is given by two spherical triangles of angles $\frac{\pi}{2}, \frac{\pi}{2}, \frac{\pi}{p}$ with common vertices at the North Pole $(0, 0, 1)$ of S^2 and at $(1, 0, 0)$, and the remaining vertices at $(\cos \frac{\pi}{p}, -\sin \frac{\pi}{p}, 0)$ and $(\cos \frac{\pi}{p}, \sin \frac{\pi}{p}, 0)$. In all these cases we take $\mathcal{D} = \mathbb{P}_1$.

This brings us to the subject of triangle groups and their relation to Gauss hypergeometric functions.

2.6 GAUSS HYPERGEOMETRIC FUNCTIONS

Many different functions are called *hypergeometric functions*. Perhaps the oldest one is the so-called classical hypergeometric function of one complex variable, also known as the *Gauss hypergeometric function*. This function, which occurs in many different branches of mathematics, satisfies a second order ordinary linear homogeneous differential equation with regular singular points at $0, 1, \infty$. In 1857, Riemann [119] showed that a function with two linearly independent branches at each of the points $0, 1, \infty$, and with suitable branching behavior at these points, necessarily satisfies a so-called hypergeometric differential equation and hence is itself a hypergeometric function. This is part of Riemann's famous "viewpoint": analytic functions are characterized by their behavior at singular points. In 1873, Schwarz [128] compiled the list of algebraic Gauss hypergeometric functions. He also established necessary conditions for the quotient of two solutions of a Gauss hypergeometric differential equation to be invertible, with inverse an automorphic function on the unit disk. In the 1880s, Picard generalized Riemann's approach to the two-variable Appell hypergeometric functions. Those functions are the subject of Chapter 7, and are related to complex two-ball quotients.

Let a, b, c be complex numbers with $c \neq 0$ and not a negative integer. Consider the differential equation

$$x(1 - x)\frac{d^2 y}{dx^2} + (c - (a + b + 1)x)\frac{dy}{dx} - aby = 0. \qquad (2.26)$$

Euler introduced the following series solution to this equation:

$$F = F(a, b, c; x) = \sum_{n=0}^{\infty} \frac{(a, n)(b, n)}{(c, n)} \frac{x^n}{n!}, \qquad |x| < 1, \qquad (2.27)$$

where for any complex number w we define $(w, n) = w(w + 1) \ldots (w + n - 1)$, $n \geq 1$, and we define $(w, 0) = 1$. Letting $A_n = \frac{(a,n)(b,n)}{(c,n)n!}$, we see that

$$\frac{A_{n+1}}{A_n} = \frac{(a + n)(b + n)}{(c + n)(1 + n)}.$$

If a or b is zero or a negative integer, then $F(a, b, c; x)$ is a polynomial in x; otherwise,

$$\frac{A_{n+1}}{A_n} \mapsto 1$$

as $n \mapsto \infty$, so the series converges for $|x| < 1$ and has $|x| = 1$ as its circle of convergence. To see that F is a solution of the differential equation, set $D = x\frac{d}{dx}$ and notice that if P is a polynomial then

$$P(D)x^n = P(n)x^n.$$

Using this and the formula for $\frac{A_{n+1}}{A_n}$, we see that

$$\left\{(a + D)(b + D) - (c + D)(1 + D)x^{-1}\right\} \left(\sum_{n=0}^{\infty} A_n x^n\right)$$

$$= \sum_{n=0}^{\infty} \{(a + n)(b + n)A_n x^n - (c + n - 1)n A_n x^{n-1}\}$$

$$= \sum_{n=0}^{\infty} (a + n)(b + n) A_n x^n - \sum_{n=0}^{\infty} (c + n)(1 + n) A_{n+1} x^n = 0,$$

which proves that $F(a, b, c; x)$ satisfies (2.26). The differential equation (2.26) can be written in the form

$$\frac{d^2 y}{dx^2} + p(x)\frac{dy}{dx} + q(x)y = 0,$$

where $p(x)$ and $q(x)$ are rational functions of x having poles only at $x = 0, 1, \infty$. These are therefore the singular points of the differential equation. They are regular singular points because $(x - \xi)p(x)$ and $(x - \xi)^2 q(x)$ are both holomorphic at $x = \xi$ when $\xi = 0$ or 1, and $p(\frac{1}{t})\frac{1}{t}$ and $q(\frac{1}{t^2})\frac{1}{t^2}$ are holomorphic at $t = 0$ (corresponding to $\xi = \infty$).

Now suppose that $\text{Re}(a), \text{Re}(c - a) > 0$. Then, the above series has an integral representation:

$$F = F(a, b, c; x) = \frac{1}{B(a, c - a)} \int_1^{\infty} u^{b-c}(u - 1)^{c-a-1}(u - x)^{-b} du$$

$$= \frac{1}{B(a, c - a)} \int_0^1 u^{a-1}(1 - u)^{c-a-1}(1 - ux)^{-b} du,$$

where, for $\text{Re}(\alpha), \text{Re}(\beta) > 0$,

$$B(\alpha, \beta) = \int_0^1 u^{\alpha-1}(1-u)^{\beta-1} du = \frac{\Gamma(\alpha)\Gamma(\beta)}{\Gamma(\alpha+\beta)}. \tag{2.28}$$

To check this integral representation, use the formula

$$(1 - ux)^{-b} = \sum_{n=0}^{\infty} \frac{(b, n)}{(1, n)} u^n x^n, \qquad |x| < 1.$$

The hypergeometric series has an analytic continuation outside its circle of convergence, $|x| = 1$, to the complex plane minus the segment $[1, \infty)$. This extended function is also called the *Gauss hypergeometric function*.

If $c, c - a - b, a - b \notin \mathbb{Z}$, then a solution of the Gauss hypergeometric differential equation linearly independent from $F(a, b, c; x)$ is given in $|x| < 1$ by the series

$$G(x) = x^{1-c} F(a + 1 - c, b + 1 - c, 2 - c; x).$$

When $\text{Re}(c - b) < 1$ and $\text{Re}(b) < 1$, the series $G(x)$ is equal, up to a constant, to

$$\int_0^x u^{b-c}(u - 1)^{c-a-1}(u - x)^{-b} du$$

in the region $|x| < 1$, as can be seen by a simple computation using the change of variables $u \mapsto \frac{x}{v}$. When one of $c, c - a - b, a - b$ lies in $\mathbb{Z}$, one of the two independent solutions around a regular singular point has a logarithmic singularity at that point. For example, when $c = 1$, the above two series solutions coincide. For $c < 1$, the function

$$\frac{1}{1-c} \left\{ x^{1-c} F(a + 1 - c, b + 1 - c, 2 - c; x) - F(a, b, c; x) \right\}$$

is a solution of the hypergeometric differential equation. The function given by

$$\lim_{c \to 1} \frac{1}{1-c} \left\{ x^{1-c} F(a + 1 - c, b + 1 - c, 2 - c; x) - F(a, b, c; x) \right\}$$

is a solution if $c = 1$ (see [25], pp. 136–138), and has the form

$$F(a, b, 1; x) \log x + F^*(a, b, 1; x),$$

where F^* is defined by the equation

$$F^*(a, b, 1; x) = \frac{\partial F}{\partial a} + \frac{\partial F}{\partial b} + 2\frac{\partial F}{\partial c}.$$

The series expansion of this function in $|x| < 1$ is computed in [25], p. 137.

Let $\mu_0 = c - b$, $\mu_1 = 1 + a - c$, $\mu_2 = b$, and $\mu_3 = 2 - (\mu_0 + \mu_1 + \mu_2)$. Then the conditions $\mathrm{Re}(a)$, $\mathrm{Re}(c - a) > 0$ become $\mathrm{Re}(\mu_1)$, $\mathrm{Re}(\mu_3) < 1$. Imposing the stronger conditions, $0 < \mu_i < 1$ for $i = 0, \ldots, 3$, the μ_i satisfy

$$\sum_{i=0}^{3} \mu_i = 2, \quad 0 < \mu_i < 1, \quad i = 0, \ldots, 3. \tag{2.29}$$

Deligne and Mostow [34] call such a quadruple a *ball quadruple*, for reasons we shall see shortly (see, also, Chapter 7). For such a quadruple with $\mu_i \in \mathbb{Q}$, $i = 0, \ldots, 3$, and for $x \in \mathbb{P}_1 \setminus \{0, 1, \infty\}$,

$$u^{-\mu_0}(u - 1)^{-\mu_1}(u - x)^{-\mu_2}du \tag{2.30}$$

is a differential form of the first kind on a smooth projective curve $X(N, x)$ with affine model

$$w^N = u^{N\mu_0}(u - 1)^{N\mu_1}(u - x)^{N\mu_2}, \tag{2.31}$$

where N is the least common multiple of the denominators of the μ_i. If $v = \frac{1}{u}$, then

$$u^{-\mu_0}(u - 1)^{-\mu_1}(u - x)^{-\mu_2}du = -v^{-\mu_3}(1 - v)^{-\mu_1}(1 - vx)^{-\mu_2}dv,$$

so μ_3 is the exponent of the integrand at $u = \infty$. Consider the six integrals

$$\int_{g}^{h} u^{-\mu_0}(u - 1)^{-\mu_1}(u - x)^{-\mu_2}du, \tag{2.32}$$

with $g, h \in \{0, 1, \infty, x\}$. These are the integral representations of six series solutions comprising two independent series solutions at each of the singular points $x = 0, 1, \infty$. For example, taking $g = 0$ and $h = x$ we obtain, up to a constant, the integral expression for $G(x)$.

We can understand these six solutions as originating in a natural S_4-symmetry permuting the elements of $\{0, 1, \infty, x\}$. Namely, consider the configuration space of four distinct points on the projective line given by the quotient

$$X(4) = \{(x_1, \ldots, x_4) \in \mathbb{P}_1^4 : x_i \neq x_j, i \neq j\}/\mathrm{Aut}(\mathbb{P}_1) \tag{2.33}$$

where $\mathrm{Aut}(\mathbb{P}_1)$ acts diagonally. There is a natural action of S_4 on $X(4)$ induced by permutation of the factors of $\mathbb{P}_1^4$. Any three distinct ordered points in $\mathbb{P}_1$ can be taken to any other three distinct ordered points in $\mathbb{P}_1$ by an element of $\mathrm{Aut}(\mathbb{P}_1)$. Therefore, there is a (noncanonical) isomorphism of $X(4)$ with $\mathbb{P}_1 \setminus \{0, 1, \infty\}$. More explicitly, consider the cross-ratio

$$V(x_1, x_2, x_3, x_4) = \frac{(x_1 - x_4)(x_2 - x_3)}{(x_1 - x_3)(x_2 - x_4)}. \tag{2.34}$$

This is defined for $x_1, x_2, x_3, x_4 \in \mathbb{P}_1$, provided that no three of these points are equal. The cross-ratio is invariant under the diagonal action of $\mathrm{Aut}(\mathbb{P}_1)$, so it is well defined on $X(4)$ and maps it bijectively onto $\mathbb{P}_1 \setminus \{0, 1, \infty\}$.

The natural action of S_4 on $X(4)$ has fixed points. Any permutation of the form $(ij)(kl)$, where the set $\{i, j, k, l\} = \{1, 2, 3, 4\}$, operates trivially. Indeed, the cross-ratio is invariant under such permutations. In $\mathbb{P}_1^4$, the points (x_1, x_2, x_3, x_4) invariant under $(ij)(kl)$ have $x_i = x_j$ and $x_k = x_l$. If, in addition, we require that no three of these coordinates be equal, then $x_i \neq x_k$. For example, corresponding to $(14)(23)$, we have points of the form (x, y, y, x), with $x, y \in \mathbb{P}_1$ and $x \neq y$. Then,

$$V(x, y, y, x) = 0,$$

and, modulo the diagonal action of $\mathrm{Aut}(\mathbb{P}_1)$ on $\mathbb{P}_1^4$, all points of the form (x, y, y, x) are equivalent. Replacing $(14)(23)$ by $(12)(34)$, we obtain the $\mathrm{Aut}(\mathbb{P}_1)$ equivalent points of the form (x, x, y, y), $x \neq y$, for which

$$V(x, x, y, y) = 1.$$

Similarly, using $(13)(24)$, we have the $\mathrm{Aut}(\mathbb{P}_1)$ equivalent points of the form (x, y, x, y), for which

$$V(x, y, x, y) = \infty.$$

Let $\overline{X(4)}$ be the quotient, by the diagonal action of $\mathrm{Aut}(\mathbb{P}_1)$, of

$$\{(x_i) \in \mathbb{P}_1^4 : x_i \neq x_j, i \neq j\} \cup \{(x, y, y, x), (x, x, y, y), (x, y, x, y), x \neq y \in \mathbb{P}_1\}.$$

Then, the cross-ratio extends to $\overline{X(4)}$ and maps it bijectively onto $\mathbb{P}_1$. The quotient of S_4 by the normal subgroup of order 4 generated by permutations of the form $(ij)(kl)$, where $\{i, j, k, l\} = \{1, 2, 3, 4\}$, is isomorphic to S_3, which has order 6.

2.7 TRIANGLE GROUPS

If the $(1 - \mu_i - \mu_j)^{-1}$ are integers for $i \neq j$, a condition we denote by **INT**, then we can relate the above hypergeometric functions to triangle groups acting discontinuously on one of the simply-connected Riemann surfaces. Namely, let

$$p = |1 - \mu_0 - \mu_2|^{-1}, \quad q = |1 - \mu_1 - \mu_2|^{-1}, \quad r = |1 - \mu_3 - \mu_2|^{-1}.$$
(2.35)

By a triangle we mean a region bounded, in spherical geometry by three great circles on the Riemann sphere, in Euclidean geometry by three straight lines, and in hyperbolic geometry (in the unit disk) by three circles orthogonal to the boundary of the unit disk. Consider the triangle with vertex angles π/p, π/q, π/r. Then, the relevant geometry depends on the angle sum as follows:

$$\frac{1}{p} + \frac{1}{q} + \frac{1}{r} > 1, \qquad \text{spherical}$$

$$\frac{1}{p} + \frac{1}{q} + \frac{1}{r} = 1, \qquad \text{Euclidean}$$

$$\frac{1}{p} + \frac{1}{q} + \frac{1}{r} < 1, \qquad \text{hyperbolic.}$$

Let T denote the interior of a triangle with angles π/p, π/q, π/r and let $\overline{T}$ denote its closure. Then, by Riemann's mapping theorem with boundary, there is a bijective and conformal map $u = u(z)$ of $\mathcal{H}$ onto T that extends continuously to $\mathbb{R}$. Here, we view T, respectively, as a simply-connected subset of $\mathbb{P}_1$ in the spherical case, of $\mathbb{C}$ in the Euclidean case, and of $\mathcal{H}$, or the unit disk, in the hyperbolic case. As above, we denote these simply-connected domains by $\mathcal{D}$. Applying a Möbius transformation if necessary, we may assume that u maps $z = 0, 1, \infty$ to the three vertices of $\overline{T}$, with respective angles π/p, π/q, π/r.

Schwarz showed that the function $u = u(z)$ satisfies the differential equation

$$\{u, z\} =: \frac{2u'u''' - 3u''^2}{2u'^2} = f(z).$$
(2.36)

The symbol $\{u, z\}$ is known as the Schwarzian derivative of u with respect to z and is invariant under the change of variables

$$u \mapsto \frac{au + b}{cu + d}, \qquad \begin{pmatrix} a & b \\ c & d \end{pmatrix} \in \mathrm{PSL}(2, \mathbb{C}).$$

Schwarz also calculated the function $f(z)$ explicitly. The result is

$$f(z) = \frac{1 - (\frac{1}{p})^2}{2z^2} + \frac{1 - (\frac{1}{q})^2}{2(1-z)^2} + \frac{1 - (\frac{1}{p})^2 - (\frac{1}{q})^2 + (\frac{1}{r})^2}{2z(1-z)}. \tag{2.37}$$

By direct calculation, we find two independent solutions of the Gauss hypergeometric differential equation (2.26) with

$$a = \frac{1}{2}\left(1 - \frac{1}{p} - \frac{1}{q} + \frac{1}{r}\right)$$

$$b = \frac{1}{2}\left(1 - \frac{1}{p} - \frac{1}{q} - \frac{1}{r}\right)$$

$$c = \left(1 - \frac{1}{p}\right).$$

Their ratio satisfies

$$\{u, z\} = \frac{1 - (\frac{1}{p})^2}{2z^2} + \frac{1 - (\frac{1}{q})^2}{2(1-z)^2} + \frac{1 - (\frac{1}{p})^2 - (\frac{1}{q})^2 + (\frac{1}{r})^2}{2z(1-z)}.$$

Schwarz showed that every regular nonconstant solution of this differential equation maps the upper half plane onto a triangle with angles $\pi/p, \pi/q, \pi/r$. This "triangle map" extends across one of the intervals $(\infty, 0), (0, 1)$, or $(1, \infty)$ to a biholomorphic map of the lower half plane $\mathcal{H}^-$ onto the image T^- of T by reflection through the corresponding side of $\overline{T}$. Changing the branch of the triangle map by going around any of the points $z = 0, 1, \infty$ changes its value, and, by the Schwarz reflection principle, the corresponding images of $\mathcal{H} \cup \mathcal{H}^-$ are nonintersecting copies of $T \cup T^-$. The change of branch corresponds to an action of the monodromy group (see §2.8) of the corresponding Gauss hypergeometric differential equation acting by Möbius transformations on $\mathcal{D}$ and provides a representation of the fundamental group of $\mathbb{P}_1 \setminus \{0, 1, \infty\}$. This monodromy group is in fact in the conjugacy class of a triangle group $\Delta = \Delta(p, q, r)$. By definition, the latter are groups of Möbius transformations determined up to conjugation by the presentation

$$\langle M_1, M_2, M_3; M_1^p = M_2^q = M_3^r = M_1 M_2 M_3 = \text{Id}\rangle. \tag{2.38}$$

The closure of a fundamental domain for the triangle group is given by $\overline{T} \cup \overline{T^-}$. If a transformation $M_i, i = 1, 2, 3$, has finite order m, it is conjugate in Δ to a rotation (in the geometry determined by the signature (p, q, r)) through $2\pi/m$ about the vertex of angle π/m of $\overline{T}$. If the order is infinite, then the transformation is conjugate to a translation. The vertices of $\overline{T}$ are, in any case, fixed points of Δ.

The list of spherical and Euclidean signatures can be computed directly. The possibilities in the spherical case (see Theorem 2.1) are

$$(2, 2, \nu)$$

$$(2, 3, 3)$$

$$(2, 3, 4)$$

$$(2, 3, 5)$$

with $2 \le \nu < \infty$, and in the Euclidean case are

$$(2, 2, \infty)$$

$$(2, 3, 6)$$

$$(2, 4, 4)$$

$$(3, 3, 3).$$

The successive images of the fundamental domain of $\Delta = \Delta(p, q, r)$ give a tesselation of $\mathcal{D}$ by triangles. In the spherical case, this tesselation consists of twice as many triangles as the order of the corresponding triangle group. In the Euclidean and hyperbolic cases it consists of an infinite number of triangles, and the corresponding triangle groups are also infinite.

The Schwarz triangle map is invertible and, by the Schwarz reflection principle, maps $T \cup T^-$ to $\mathcal{H} \cup \mathcal{H}^-$. The function can be continued analytically onto all of $\mathcal{D}$ to yield a mapping

$$j : \mathcal{D} \to \mathbb{P}_1 \setminus \{\text{infinite ramification points}\}$$

automorphic with respect to Δ and ramified over 0 with order p (if $p < \infty$), over 1 with order q (if $q < \infty$), and over ∞ with order r (if $r < \infty$).

The case $(p, q, r) = (2, 3, \infty)$ corresponds to the elliptic modular group, that is, a representative of the conjugacy class of $\Delta(2, 3, \infty)$ is given by $\mathrm{SL}(2, \mathbb{Z})$, the group of 2×2 matrices with rational integer entries and determinant 1. It is generated by the two transformations on $\mathcal{H}$ given by $S : z \mapsto -1/z$ and $T : z \mapsto z + 1$. The function j has a Fourier expansion that may be normalized as

$$j(z) = e^{-2\pi i z} + 744 + \sum_{n=1}^{\infty} a_n e^{2\pi i n z},$$

with the a_n positive integers determined by the first two terms in the above series for j.

For proofs of most of the results of this section, see [25]. The discussion of the Schwarz triangle maps does not in fact require the **INT** assumption. For the more general result due to Schwarz, see [25], pp. 134–135. Yet, as Schwarz also showed, the INT assumption does ensure that the action of the corresponding triangle group is discontinuous. Knapp [77] lists the hyperbolic triangles whose angles are not of the form π/p, π/q, π/r, for $2 \leq p, q, r \leq \infty$ positive integers, but whose triangle groups are still discrete in PSL$(2, \mathbb{R})$ (see also [108]).

We mention here the principal congruence subgroups of $\Gamma = \mathrm{SL}(2, \mathbb{Z})$, which play an important role in arithmetic questions. For every integer $N \geq 1$, we define the principal congruence subgroup of level N to be

$$\Gamma[N] = \left\{ \begin{pmatrix} a & b \\ c & d \end{pmatrix} \in \Gamma \mid a \equiv d \equiv 1, b \equiv c \equiv 0 \ (\mathrm{mod}\ N) \right\}.$$

The group $\Gamma[N]$ is normal in Γ as it is the kernel of reduction mod N. The quotient group $\Gamma/\Gamma[N]$ for $N \geq 3$ has order

$$\mu(N) = \frac{1}{2} \cdot N^3 \cdot \prod_{p|N}(1 - 1/p^2),$$

where the product is over the prime divisors of N (see [90], Chapter 6). The non-compact quotient space $\Gamma[N]\backslash\mathcal{H}$ is called a (principal) modular curve of level N. If $N = p$ is itself prime, then $\Gamma/\Gamma[p]$ is isomorphic to PSL$(2, \mathbb{F}_p)$, which is simple for $p \geq 5$. For example, for $p = 5$ we have a simple group of order 60 isomorphic to the alternating group A_5, and for $p = 7$ a simple group of order 168 usually denoted by G_{168}. The group A_5 is also known as the icosahedral group, since it is isomorphic to the subgroup of rotations in $\mathbb{R}^3$, which are symmetries of the icosahedron. The latter can be inscribed in the sphere S^2, with its vertices on the surface of the sphere. Clearly, the transformation T has order 5 in A_5 and the transformations S and ST have orders 2 and 3 respectively, as they do in Γ. In fact, the group A_5 is isomorphic to the spherical triangle group of signature $(2, 3, 5)$ in the list given above. It has a conjugacy class of fifteen elements of order 2, corresponding to the fifteen edge pairs; two classes of ten elements of order 3, arising from the ten face midpoint pairs; and four classes of six elements of order 5, arising from the six vertex pairs. These pairs on the icosahedron are matched by antipodal identification $S^2/\{\pm 1\}$. The thirty edge pairs on the icosahedron give rise to fifteen great circles on S^2, which become fifteen projective lines in $\mathbb{P}_2(\mathbb{R})$. These lines form the icosahedral arrangement, to which we return in Chapter 5. The action of A_5 descends to $\mathbb{P}_2(\mathbb{R})$, the lines of the icosahedral arrangement being fixed lines of the fifteen involutions of A_5. The case $p = 7$ was studied by Klein [76]. In the quotient G_{168}, the transformation T above clearly has order 7, and the transformations S and ST have orders 2 and 3 respectively, as they do in Γ. The conjugacy class of S in G_{168} gives rise to 21 elements of order 2. There are two further conjugacy classes of elements of order 4, each having twenty-one elements; a conjugacy class of ST consisting of fifty-six elements of order 3; and two conjugacy classes, one for T, of elements of order 7, each with twenty-four elements. The twenty-one elements of order 2 correspond to the twenty-one lines of the Klein configuration (see Chapter 5). As the Euler number of the modular curve $\Gamma[7]\backslash\mathcal{H}$ is

$$\mu(7)\left(-1 + \frac{1}{2} + \frac{1}{3} + \frac{1}{7}\right) = -4,$$

it has genus 3. Therefore, the action of G_{168} on the space of differential forms of the curve gives rise to a three-dimensional linear representation and an induced action on the projective plane.

2.8 THE HYPERGEOMETRIC MONODROMY GROUP

Again consider the differential equation for the Gauss hypergeometric function (2.26). Let $x_0 \in \mathbb{P}_1 \setminus \{0, 1, \infty\}$ and let y_1, y_2 be two linearly independent solutions of (2.26) in a neighborhood of x_0. If we analytically continue y_1 and y_2 around a closed curve C in $\mathbb{P}_1 \setminus \{0, 1, \infty\}$ that starts and ends at x_0, the functions remain linearly independent solutions. Since y_1 and y_2 span the solution space, there is a nonsingular matrix

$$M(C) = \begin{pmatrix} a & b \\ c & d \end{pmatrix}$$

with coefficients in $\mathbb{C}$ such that y_1 becomes $ay_1 + by_2$ and y_2 becomes $cy_1 + dy_2$ upon analytic continuation around C. A different choice of base point x_0 would yield a matrix in the same $\mathrm{GL}_2(\mathbb{C})$ conjugacy class as $M(C)$. Denoting by $C_1 \circ C_2$ the composition of two closed curves with endpoints at x_0, we have

$$M(C_1 \circ C_2) = M(C_1)M(C_2),$$

with matrix multiplication on the right-hand side of this equation. If C_1 can be continuously deformed in $\mathbb{P}_1 \setminus \{0, 1, \infty\}$ into C_2, with x_0 fixed, then $M(C_1) = M(C_2)$. Let $\pi_1(\mathbb{P}_1 \setminus \{0, 1, \infty\}, x_0)$ denote the group of homotopy equivalence classes of curves starting and ending at the base point x_0. This is the fundamental group defined in Chapter 1, §3. From the above remarks, we see that we have a homomorphism

$$M : \pi_1(\mathbb{P}_1 \setminus \{0, 1, \infty\}, x_0) \to GL_2(\mathbb{C}),$$

called the *monodromy representation*, associated to the differential equation. The monodromy group of (2.26) is defined as the image of the monodromy representation. The projective monodromy group of (2.26) is defined as the image of the monodromy group under the natural map $GL_2(\mathbb{C}) \to PGL_2(\mathbb{C})$. If we change the base point x_0 or the choice of basis of the solution space of (2.26), then we conjugate M by an element of $GL_2(\mathbb{C})$. The conjugacy class of the monodromy group and the projective monodromy group are uniquely determined by (2.26).

The signature (p, q, r) of a triangle group determines its representation, up to conjugacy, in the group of Möbius transformations. The image of the monodromy representation when

$$p = |1 - \mu_0 - \mu_2|^{-1}, q = |1 - \mu_1 - \mu_2|^{-1}, r = |1 - \mu_3 - \mu_2|^{-1}$$

is $\Delta(p, q, r)$, as given in (2.38).

Chapter Three

Complex Surfaces and Coverings

We now turn our attention from Riemann surfaces to surfaces of two complex dimensions. Once again, we will study coverings branched along subvarieties of codimension 1. Since Riemann surfaces have complex dimension 1, divisors on them are merely finite sums of points with integer coefficients. In the case of two complex dimensions, however, divisors are finite sums of one-dimensional subvarieties with integer coefficients. We are now faced with a complication: two or more of these subvarieties may have a point in common. We begin our study of branched coverings in §3.1, where we suppose the branch locus consists of subvarieties that intersect transversally, that is, no more than two subvarieties meet at a common point. In order to treat coverings branched over subvarieties with more than double intersection points, we use the technique of blowing up a point to reduce the problem to the case of transverse intersections. This blowing-up process is discussed in some detail at the end of the chapter in §3.5. In §3.2, we introduce the Chern numbers of a complex surface. This enables us to define the proportionality deviation of a complex surface, denoted by Prop, in §3.3, and to study its behavior with respect to finite covers. The vanishing of Prop is intimately related to the existence of finite covers of line arrangements that are ball quotients. This is the main topic of this book and is the subject of Chapters 5 and 6. We show how solutions of Prop=0 arise from the complete quadrilateral arrangement. Finally, we discuss the signature of a complex surface in §3.4.

3.1 COVERINGS BRANCHED OVER SUBVARIETIES WITH TRANSVERSE INTERSECTIONS

Let X denote a complex surface, that is, a complex manifold of complex dimension 2. Unless otherwise stated, we assume from now on that our complex surfaces are compact, connected, and algebraic. Let D_i, $i \in I$ for some finite index set I, be a set of smooth complex one-dimensional irreducible subvarieties. These have the underlying structure of a compact Riemann surface. In a neighborhood of any point $p \in D_i$, we can choose local complex coordinates (u, v) on X such that D_i is given locally by the equation $u = 0$, and u is called a normal coordinate to D_i at p. Assume that any two distinct subvarieties in the system intersect transversally. That is, for any $i \neq j$ with $i, j \in I$ and any $p \in D_i \cap D_j$, we have a normal crossing at p, that is, we can find local coordinates (u, v) at p such that D_i is given locally by $u = 0$ and D_j is given locally by $v = 0$. The cardinality of $D_i \cap D_j$, $i \neq j$, is therefore finite and is by definition the *intersection number* $D_i \cdot D_j$. We shall say more about the definition of intersection numbers in §3.2. In addition, we assume that no more than two of the D_i intersect at one point; in other words, the union $\cup_i D_i$ has only ordinary double points.

We shall deal mainly with *good covers* as given by the following (compare with [75], p. 150, [137], and [138]):

Definition 3.1 *Let Y be a complex surface that is a finite covering*

$$\pi : Y \to X$$

of X. Therefore, the map π is holomorphic and surjective, and above any point of X there are only a finite number of points of Y. We suppose that π is branched along a system $\{D_i\}_{i \in I}$ of one-dimensional subvarieties of X intersecting transversally. The covering is defined to be a good covering if, in addition, there are integers $N \geq 1$ and $b_i \geq 2$, $i \in I$, such that

(i) *for $i \in I$, we have $b_i \mid N$ and there are N/b_i points of Y over each point of $D_i \setminus \cup_{j \neq i} D_j \cap D_i$: centered at each such point q of Y, there are local coordinates (s, t) such that $u = s^{b_i}$, $v = t$ are local coordinates centered at $\pi(q)$, with u a normal coordinate to D_i at $\pi(q)$; the map π is given locally by the quotient of an open neighborhood of q by the action of $\mathbb{Z}/b_i\mathbb{Z}$ by $(s, t) \mapsto (\exp(2\pi i m/b_i)s, t)$ for $m \in \mathbb{Z}/b_i\mathbb{Z}$;*

(ii) *for $i, j \in I$, $i \neq j$ and $D_i \cap D_j \neq \phi$, we have $b_i b_j \mid N$ and there are $N/b_i b_j$ points over each point of $D_i \cap D_j$; centered at each such point q of Y, there are local coordinates (s, t) such that $u = s^{b_i}$, $v = t^{b_j}$ are local coordinates centered at $\pi(q)$, with u a normal coordinate to D_i at $\pi(q)$ and v a normal coordinate to D_j at $\pi(q)$; the map π is given locally by the quotient of an open neighborhood of q by the action of $(\mathbb{Z}/b_i\mathbb{Z}) \times (\mathbb{Z}/b_j\mathbb{Z})$ by $(s, t) \mapsto (\exp(2\pi i m/b_i)s, \exp(2\pi i n/b_j)t)$ for $m \in \mathbb{Z}/b_i\mathbb{Z}$, $n \in \mathbb{Z}/b_j\mathbb{Z}$;*

(iii) *over the points not appearing in (i) and (ii) there are N points of Y, and N is called the degree of the covering; at any such point of Y, the map π is locally biholomorphic.*

Consider the following example of a covering of the complex projective plane $\mathbb{P}_2$ by itself that has degree n^2. Let $(w_0 : w_1 : w_2)$ and $(z_0 : z_1 : z_2)$ be projective coordinates on $\mathbb{P}_2$ and consider the map $w_i = z_i^n$, $i = 0, 1, 2$. The subvarieties $w_i = 0$, $i = 0, 1, 2$, on X intersect transversally at one point.

In terms of affine coordinates on the open subset $w_0 \neq 0$ of $\mathbb{P}_2$, we have a covering branched along $u = w_1/w_0 = 0$, $v = w_2/w_0 = 0$ on which the functions $^n\sqrt{u}$, $^n\sqrt{v}$ become single-valued.

We can compute the Euler number of a good covering Y using the properties of the Euler number given in Chapter 1, §1.1. Namely,

$$e(Y) = N \left(e(X) - \sum_i e(D_i) + \frac{1}{2} \sum_{i \neq j} D_i \cdot D_j \right)$$

$$+ N \sum_i \frac{1}{b_i} \left(e(D_i) - \sum_{j \neq i} D_i \cdot D_j \right) + \frac{N}{2} \sum_{i \neq j} \frac{1}{b_i b_j} D_i \cdot D_j. \qquad (3.1)$$

In the first part of the sum, we compute the contribution to the Euler number of the complement of the ramification locus and, in the second part, we take into account the contribution from the ramification locus. Let $x_i = 1 - \frac{1}{b_i}$. Then, the above formula for the Euler number takes the simple form

$$\frac{e(Y)}{N} = e(X) - \sum_i x_i e(D_i) + \frac{1}{2} \sum_{i \neq j} x_i x_j D_i \cdot D_j. \qquad (3.2)$$

Compare this to the analogous one-dimensional formula for a covering Y_1 of a Riemann surface X_1 of degree N ramifying with degree b_i over points P_i, with N/b_i points of Y_1 over P_i (see Chapter 2, §2.4):

$$\frac{e(Y_1)}{N} = e(X_1) - \sum_i x_i e(P_i). \qquad (3.3)$$

In higher dimensions we can check that, under suitable assumptions, the expression for the Euler number resembles a Taylor formula. This might be expected on consideration of the one-dimensional and two-dimensional cases.

3.2 DIVISOR CLASS GROUP AND CANONICAL CLASS

Recall the notion of divisor from Chapter 1, §1.4. The group of divisors on X is the abelian group generated freely by the irreducible analytic subvarieties of X of complex dimension 1, called the *prime divisors*. These subvarieties are therefore curves on X, which are not necessarily smooth. Let $D = \sum_i m_i C_i$ be a divisor, where the m_i are integers and the C_i are prime divisors. There is an open covering of X such that, for every open set U in the covering, the intersection $C_i \cap U$ is the locus of the equation $f_U^i = 0$, where f_U^i generates a prime ideal in the ring of regular functions on U. Therefore, on U, the divisor D is given by the locus of zeros and poles of $f_U = \prod_i (f_U^i)^{m_i}$, counted with multiplicity, and this is independent of the chosen open covering. Therefore, for some open covering of X, and for every open set U in the covering, the divisor is given on U by the zeros and poles, counted with multiplicity, of a meromorphic function f_U on U, which is not identically zero on U. Moreover, on the intersection $U \cap V$ of two sets U and V belonging to the covering, the function f_U/f_V is holomorphic and nonzero. Addition and subtraction of divisors corresponds to multiplication and division of their local functions. A nonzero global meromorphic function f on X determines a divisor, (f), by letting $f_U = f$ for every U. We say that two divisors D_1 and D_2 are *linearly equivalent* if there is a global meromorphic function f on X such that $D_1 - D_2 = (f)$. The corresponding equivalence classes form the group $\mathrm{Div}(X)$ of divisor classes on X. If $\pi : Y \to X$ is a surjective map between complex surfaces, then the pullback on functions induces a homomorphism $\pi^* : \mathrm{Div}(X) \to \mathrm{Div}(Y)$.

Let ω be a nonzero meromorphic differential 2-form on X. In terms of local complex coordinates (z_1, z_2), it has the form

$$\omega(z_1, z_2) = a(z_1, z_2)dz_1 \wedge dz_2,$$

where $a(z_1, z_2)$ is a meromorphic function. For an arbitrary complex surface, such a form need not exist. However, all algebraic surfaces have such forms. We use these forms to define the canonical class and to derive properties of the canonical class under ramification. Nonetheless, the theory applies for all complex surfaces. Under a change of complex coordinates, $dz_1 \wedge dz_2$ is multiplied by the determinant of the Jacobian of the coordinate change. Therefore, we may use the meromorphic function $a(z_1, z_2)$ to locally describe a divisor. The quotient of any two meromorphic differential 2-forms is a meromorphic function, and so they define linearly equivalent divisors. Therefore, in this way we obtain a well-defined divisor class, which we call the canonical class $K = K_X$ of X (see also Chapter 1, §1.4).

Let D be a representative of a divisor class on X. Then D is also an oriented two-dimensional real cycle on the real four-dimensional manifold X, and determines an element $[D]$ of the homology group $H_2(X, \mathbb{Z})$. As linearly equivalent divisors give rise to homologous elements of $H_2(X, \mathbb{Z})$ (see [55]), $[D]$ depends only on the divisor class of D. The homology class $[K]$ is called the *canonical homology class*. On a four-dimensional oriented real compact manifold X, we can always choose a smooth representative for every class in $H_2(X, \mathbb{Z})$. The intersection number of two homology classes is defined by moving their smooth representatives topologically so that they are in transverse positions, and then using the definition of intersection number for transversally intersecting cycles. Namely, let A and B be two smooth oriented real two-dimensional submanifolds of X representing elements α and β of $H_2(X, \mathbb{Z})$. We move A and B topologically until they intersect transversally. At each intersection point, their orientations together generate either a positive or a negative orientation in the real four-dimensional tangent space at the point. The intersection number of α and β is then

$$\alpha \cdot \beta = \sum_+ 1 - \sum_- 1,$$

where $\sum_+$ denotes the sum over the intersection points where A and B give a positive orientation, and $\sum_-$ the sum over the intersection points where A and B give a negative orientation. Consider the example $X = \mathbb{P}_2$ with projective coordinates $(z_0 : z_1 : z_2)$ and divisors given by the lines $z_1 = 0$ and $z_2 = 0$. Let real coordinates (x_1, y_1) and (x_2, y_2) be given by $z_1 = x_1 + iy_1$ and $z_2 = x_2 + iy_2$, and consider the orientation determined by this complex structure, that is, by the order x_1, y_1, x_2, y_2. The intersection number of the divisors $z_1 = 0$ and $z_2 = 0$ is then 1. If we orient $\mathbb{P}_2$ in the opposite way, their intersection number becomes -1. By Poincaré duality [49], p. 53, we have a natural isomorphism between $H_2(X, \mathbb{Z})$ and the cohomology group $H^2(X, \mathbb{Z})$. The image of $[D]$ under this isomorphism is the first Chern class $c_1(D)$ (see also Chapter 1, §1.4). The intersection product in homology becomes the cup product in cohomology, composed with integration over the manifold, that is, with pairing with the fundamental homology class.

Let A be a smooth subvariety of complex codimension 1 with homology class $\alpha \in H_2(X, \mathbb{Z})$. The normal bundle $\mathcal{N}_{A/X}$ to A in X is given by the quotient bundle $T_X|_A / T_A$, where for M a manifold, T_M denotes its tangent bundle. Consider any continuous section of a bundle with isolated zeros and poles. The number of zeros and poles of this section, counted with multiplicity (which is negative for the poles), is called the *degree of the bundle*. The self-intersection $\alpha \cdot \alpha$ is the degree of $\mathcal{N}_{A/X}$. We can assume that the section of $\mathcal{N}_{A/X}$ used to compute the degree is differentiable and gives a smooth differentiable submanifold homologous to A and intersecting A transversally.

In complex geometry, two transversally intersecting smooth subvarieties D_1 and D_2 of codimension 1 always define a positive orientation at any intersection point. Their intersection number $[D_1] \cdot [D_2]$ is therefore simply the number of intersection points. Notice that on a complex surface X, this means an irreducible curve with negative self-intersection number is the only irreducible curve in its homology class.

There is an algebraically defined intersection pairing on $\mathrm{Div}(X)$ that agrees with the topological intersection form under the induced map from $\mathrm{Div}(X)$ to $H_2(X, \mathbb{Z})$. In fact, there is a unique pairing on the divisor group $\mathrm{Div}(X)$ of X,

$$\mathrm{Div}(X) \times \mathrm{Div}(X) \mapsto \mathbb{Z}$$

$$(D_1, D_2) \mapsto D_1 \cdot D_2,$$

such that

> a) if D_1 and D_2 are smooth subvarieties of codimension 1 that meet transversally, then $D_1 \cdot D_2 = \operatorname{card}(D_1 \cap D_2)$,
> b) $D_1 \cdot D_2 = D_2 \cdot D_1$
> c) $(D_1 + D_2) \cdot D = D_1 \cdot D + D_2 \cdot D$
> d) if D_1 and D_2 are linearly equivalent, then $D_1 \cdot D = D_2 \cdot D$ for every D.

The first step in the construction of this pairing is the definition of the local intersection of two curves with no component in common. Suppose that D_1 and D_2 are two divisors with no component of dimension 1 in common and that they are defined by holomorphic local functions. Let p be a point in $D_1 \cap D_2$. Then there are functions f and g in the local ring $\mathcal{O}_{X,p}$ of holomorphic functions at p giving, respectively, the local equations of D_1 and D_2. We define the local intersection number $I_p(D_1, D_2)$ to be the complex dimension of the space $\mathcal{O}_{X,p}/(f, g)$ where (f, g) is the ideal generated by f and g. If $p \notin D_1 \cap D_2$, we let $I_p(D_1, D_2) = 0$. Then,

$$D_1 \cdot D_2 = \sum_p I_p(D_1, D_2) \geq 0.$$

The divisors D_1, D_2 meet transversally at p if and only if $I_p(D_1, D_2) = 1$, as then (f, g) is the maximal ideal of $\mathcal{O}_{X,p}$. In this way, we recover property (a) of the pairing. We have to show that there is a unique symmetric bilinear pairing from $\operatorname{Div}(X)$ to $\mathbb{Z}$, which factors through linear equivalence and agrees with the above definition for curves meeting transversally. For a proof of this, see, for example, [44],[51], and [130].

By way of example, consider a smooth curve C and the diagonal

$$\Delta = \{(x, x) \mid x \in C\}$$

in $C \times C$. The Euler number of C equals the self-intersection number of Δ. Namely, the normal bundle of Δ in $C \times C$ is isomorphic to the tangent bundle of C. The number of zeros of a vector field on C, counted with multiplicity, equals the Euler number of C by the Poincaré–Hopf theorem. Hence, $\Delta \cdot \Delta = 2 - 2g$, which is negative when $g \geq 2$, so that Δ is then isolated in its homology class. Therefore, the automorphism group of C is discrete. In fact, it is finite.

Returning to the canonical class K, the first Chern class $c_1(X)$ of X is defined as $-c_1(K)$. Therefore, we have

$$c_1^2(X) = [K] \cdot [K] = K \cdot K. \tag{3.4}$$

The Euler number $e(X)$ is also denoted by $c_2(X)$. We therefore have the two Chern numbers, $c_1^2(X)$ and $c_2(X)$.

In complex dimension 1, we define, in an analogous way, the canonical divisor K using the meromorphic 1-forms. It is a linear combination of points and has degree $2g - 2$, where g is the genus of the underlying Riemann surface. Here, the Euler number $2 - 2g$ is the only Chern number.

Let us see how the canonical class behaves under covering maps. Let X and Y be smooth compact algebraic surfaces. Let $\pi : Y \mapsto X$ be a good covering as defined in Definition 3.1 of §3.1. We retain the notation of that definition. Let D be the divisor that is the locus of vanishing of the determinant of the Jacobian of π on Y. Then, looking at the meromorphic 2-forms of X and Y, we have

$$K_Y = \pi^* K_X + D. \tag{3.5}$$

In order to express this formula in terms of the branching locus D_i on X, we use rational coefficients, and obtain a two-dimensional Hurwitz formula,

$$K_Y = \pi^* K_X + \sum_i \frac{b_i - 1}{b_i} \pi^* D_i$$
$$= \pi^* (K_X + \sum_i x_i D_i), \tag{3.6}$$

where, as before, $x_i = 1 - \frac{1}{b_i}$. To understand this formula, recall that the lifting π^* of divisors corresponds to the lifting of the local meromorphic functions. Over the points of $D_i \setminus \cup_{i \neq j} D_i \cap D_j$, we have a local map $u = t^{b_i}$, $v = s$, with $du \wedge dv = b_i t^{b_i - 1} dt \wedge ds$. The lift of D_i is given by $t^{b_i} = 0$.

Consequently,

$$K_Y \cdot K_Y = N \left(K_X + \sum_i x_i D_i \right)^2. \tag{3.7}$$

In (3.7), we used the cohomological relations

$$(\pi^* \alpha \cup \pi^* \beta)[Y] = \pi^* (\alpha \cup \beta)[Y] = (\alpha \cup \beta)[\pi_* Y] = (\alpha \cup \beta) N[X]$$

for $\alpha, \beta \in H^2(X, \mathbb{Z})$.

Consider, for example, the complex projective plane $\mathbb{P}_2$. It has Euler number $e(\mathbb{P}_2) = 3$ since it is the disjoint union of a copy of $\mathbb{C}^2$ and a copy of $\mathbb{P}_1$, viewed as the projective line at ∞. Its divisor class group is generated by a projective line l, corresponding, by Poincaré duality, to an element $h \in H^2(\mathbb{P}_2)$, where h^2 pairs with the fundamental class to give 1. Choosing homogeneous coordinates $[u_0 : u_1 : u_2]$ on $\mathbb{P}_2$, we let $z_1 = u_1/u_0, z_2 = u_2/u_0$ be affine coordinates on $u_0 \neq 0$ and we let $w_1 = u_0/u_1, w_2 = u_2/u_1$ be affine coordinates on $u_1 \neq 0$. Then, the 2-form given on $u_0 \neq 0$ by $dz_1 \wedge dz_2$ has a pole of order 3 at $w_1 = 0$, since

$$dz_1 \wedge dz_2 = -w_1^{-3} dw_1 \wedge dw_2.$$

Hence, the canonical class has representative $K = -3l \simeq -3h$, and K has self-intersection $K^2 = 9$. Recall that $K^2 = c_1^2(\mathbb{P}_2)$ and that we have $c_2(\mathbb{P}_2) = e(\mathbb{P}_2) = 3$; therefore,

$$c_1^2(\mathbb{P}_2) = 3c_2(\mathbb{P}_2) = 9.$$

3.3 PROPORTIONALITY

The equality $c_1^2 = 3c_2$, satisfied by $\mathbb{P}_2$, will occupy us for a great deal of the remainder of this book, especially in Chapters 5 and 6. This comes from its relation to ball quotients. The complex 2-ball $B = B_2$ is the bounded homogeneous domain contained in $\mathbb{P}_2$ that is given in projective coordinates by

$$|z_1|^2 + |z_2|^2 < |z_0|^2.$$

The group of automorphisms of B, which act on the projective coordinates by fractional linear transformations, is denoted by $PU(2, 1)$. In fact, $PU(2, 1)$ is the group of all biholomorphic automorphisms acting on B (see [78], §4A.1, p. 16). It is naturally contained in the automorphism group $PGL(3, \mathbb{C})$ of $\mathbb{P}_2$. In 1956, F. Hirzebruch [54] proved a theorem, of which the following is a special case.

Theorem 3.1 *Assume that*

    (i) *Γ is a discrete subgroup of $PU(2, 1)$,*
    (ii) *Γ operates freely, that is, an element that is not the identity has no fixed point,*
    (iii) *B/Γ is compact.*

For such a ball quotient, the Chern numbers are proportional to those of $\mathbb{P}_2$. Therefore, we have the following equation:

$$c_1^2(B/\Gamma) = 3c_2(B/\Gamma).$$

The quotient complex surface B/Γ is algebraic, according to a result of Kodaira.

Yoichi Miyaoka and Shin-Tung Yau independently proved that $c_1^2 \leq 3c_2$ for surfaces of general type (see Chapter 4, §§4.2, 4.3). Ball quotients satisfying the conditions of Theorem 3.1 are of general type. By a famous result of Yau, they are characterized by equality in the Miyaoka–Yau inequality. This motivates the following definition.

Definition 3.2 *The proportionality deviation of a complex surface Y is defined by the following formula:*

$$\mathrm{Prop}(Y) =: 3c_2(Y) - c_1^2(Y). \tag{3.8}$$

Our goal is to calculate Prop for ramified covers and to find examples where it vanishes. This is the main objective of Chapter 5.

Recall the generalization of the Hurwitz formula given by (3.2). Rewriting it in terms of the second Chern number, we have

$$c_2(Y)/N = c_2(X) - \sum_i x_i e(D_i) + \frac{1}{2} \sum_{i \neq j} x_i x_j D_i \cdot D_j. \tag{3.9}$$

Expressing the self-intersection formula (3.7) for the canonical class in terms of the first Chern number, we have

$$c_1^2(Y)/N = (K_X + \sum_i x_i D_i)^2 = c_1^2(X) + 2 \sum_i x_i K_X \cdot D_i$$
$$+ \sum_i x_i^2 D_i \cdot D_i + \sum_{i \neq j} x_i x_j D_i \cdot D_j. \tag{3.10}$$

In order to simplify this formula, we use the following.

Adjunction formula: If D is a smooth submanifold of X of dimension 1, then

$$e(D) = -K_X \cdot D - D \cdot D. \tag{3.11}$$

For a proof of the adjunction formula, see [51], p. 361. By way of example, let D be a smooth curve of degree n in $\mathbb{P}_2$. That is, D is given by a homogeneous polynomial of degree n in homogeneous coordinates of $\mathbb{P}_2$, not all of whose derivatives vanish simultaneously. Then, as a divisor, D is equivalent to nl, where l is a projective line in $\mathbb{P}_2$. Therefore, $D \cdot D = n^2$ and $K_{\mathbb{P}_1} \cdot D = -3n$. By the adjunction formula, we have $e(D) = 3n - n^2$, and we also know that $e(D) = 2 - 2g$, where g is the genus of the curve. Therefore, $3n - n^2 = 2 - 2g$, which implies

$$g = \frac{(n-1)(n-2)}{2}.$$

Theorem 3.2 *With the notations of §3.1, for good coverings*

$$\pi : Y \to X$$

of degree N, the proportionality deviation is given by

$$\frac{1}{N}\text{Prop}(Y) = \frac{1}{N}\left(3c_2(Y) - c_1^2(Y)\right) = 3c_2(X) - c_1^2(X)$$

$$+ \sum_i x_i \left(-e(D_i) + 2D_i \cdot D_i\right) + \frac{1}{2}\sum_{i \neq j} x_i x_j D_i \cdot D_j - \sum_i x_i^2 D_i^2. \quad (3.12)$$

Proof. The formula (3.12) is obtained by combining (3.9), (3.10), and (3.11). □

If the D_i are k *lines* in general position in $\mathbb{P}_2$, then $e(D_i) = 2$, $D_i \cdot D_i = 1$, and $D_i \cdot D_j = 1$, $i \neq j$. Therefore, from (3.12), we obtain

$$\frac{1}{N}\text{Prop}(Y) = \frac{1}{N}(3c_2(Y) - c_1^2(Y)) = \frac{1}{2}\sum_{i \neq j} x_i x_j - \sum_i x_i^2.$$

Of course, $\frac{1}{2} \leq x_i \leq 1$ for $i = 1, \ldots, k$. For $k = 3$, the far right-hand side of the preceding expression equals

$$S = -\frac{1}{2}\{(x_1 - x_2)^2 + (x_2 - x_3)^2 + (x_1 - x_3)^2\},$$

and, as $|x_i - x_j| \leq \frac{1}{2}$ for all i, j, we have $S \geq -\frac{3}{8}$. Notice that any vector of the form (x, x, x) is a solution of $S = 0$. We now proceed inductively, writing

$$\frac{1}{N}\text{Prop}(Y) = \frac{1}{2}\sum_{i,j=1,\ldots,k-1; i \neq j} x_i x_j - \sum_{i \neq k} x_i^2 + x_k(x_1 + \ldots + x_{k-1} - x_k).$$

Now, as $k \geq 4$,

$$x_k(x_1 + \ldots + x_{k-1} - x_k) \geq x_k \left(\frac{k-1}{2} - x_k\right) \geq \frac{1}{2}.$$

Hence, $\text{Prop}(Y) > 0$ for all $k \geq 4$. For $k = 3$, the example following Definition 3.1 shows $\text{Prop}(Y) > 0$ is not valid.

The example of the complete quadrilateral

An arrangement that will be important in our study of Appell's hypergeometric function in Chapter 6, and that gives rise to examples where Prop does vanish, is the complete quadrilateral. It is the arrangement of six lines having four triple intersection points, no three of which are collinear (see Figure 1). Any four points with this property are equivalent up to a projective transformation.

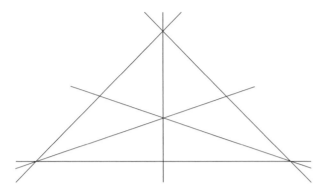

Figure 3.1 The complete quadrilateral

The six lines are the $k = \binom{4}{2} = 6$ ways of connecting these four points by lines. This arrangement has three double and four triple intersection points. Any three of its lines not having a common triple point give an affine coordinate system on an open subset of $\mathbb{P}_2$, and, in suitable projective coordinates $(z_0 : z_1 : z_2)$, the arrangement is given by

$$z_0 z_1 z_2 (z_2 - z_1)(z_2 - z_0)(z_0 - z_1) = 0.$$

The discussion in this chapter assumed that the branching locus consists of divisors with normal crossings. This is certainly not the case for the complete quadrilateral. In order to apply the considerations of the present chapter, we construct a new algebraic surface by blowing up the triple points of the complete quadrilateral. For a more complete discussion of this blowing-up process, see §3.5. Consider a triple intersection point and assume it is at the origin. Roughly speaking, when we blow up the origin, we replace each neighboring point not at the origin by the line joining it with the origin, and replace the origin itself by the family of lines through the origin. Therefore, in blowing up $\mathbb{P}_2$ at the four triple points of the complete quadrilateral, which we index by $\{0i\}$, $i = 1, \ldots, 4$, we replace them by projective lines. We therefore obtain a new surface X with a total of ten divisors $D_{\alpha\beta}$, indexed by $\alpha, \beta \in \{0, 1, \ldots, 4\}$. Six of these divisors are transforms of the original lines of the arrangement and four of them are the divisors coming from the blown-up points. For example, the divisor D_{12} is the proper transform of the line passing through the points 03 and 04, whereas D_{0i} for $i = 1, \ldots, 4$ is the divisor obtained by blowing up the point $0i$. The ten divisors have only transverse intersection points. There are fifteen such points. The intersection numbers for two divisors, $D_{\alpha\beta}$ and $D_{\gamma\delta}$, $\{\alpha\beta\} \neq \{\gamma\delta\}$, are given by

$$D_{\alpha\beta} \cdot D_{\gamma\delta} = 1 \quad \text{if} \quad \{\alpha\beta\} \cap \{\gamma\delta\} = \phi \text{ and}$$

$$D_{\alpha\beta} \cdot D_{\gamma\delta} = 0 \quad \text{if} \quad \{\alpha\beta\} \cap \{\gamma\delta\} \neq \phi,$$

and for $\{\alpha\beta\} = \{\gamma\delta\}$ by

$$D_{\alpha\beta} \cdot D_{\alpha\beta} = -1.$$

For each of the ten divisors D, we have $e(D) = 2$. Notice that for each of the original lines in the arrangement on $\mathbb{P}_2$, the self-intersection is $+1$. This self-intersection goes down by 1 on X for each blown-up point lying on the line. Therefore, we end up with self-intersection -1: again, we refer to our discussion of blow-ups in §3.5. When we blow up $\mathbb{P}_2$ to obtain X, each blown-up point augments c_2 by $+1$, as a point becomes a copy of S^2, whereas the extra curve with self-intersection -1 decreases c_1^2 by -1. Hence, we have

$$3c_2(X) - c_1^2(X) = 16.$$

Using (3.12) of Theorem 3.2, we deduce that a good covering Y of X of degree N, as given by Definition 3.1 of §3.1, satisfies

$$\frac{1}{N}\text{Prop}(Y) = 16 + \sum x_{\alpha\beta}(-4) + \sum x_{\alpha\beta}^2 + \frac{1}{2}\sum x_{\alpha\beta}\Big(\sum_{\{\alpha\beta\}\neq\{\gamma\delta\}} x_{\gamma\delta}D_{\alpha\beta} \cdot D_{\gamma\delta}\Big).$$

$$(3.13)$$

We now exhibit a formal solution to $\text{Prop} = 0$. Namely, let $\mu_0, \mu_1, \ldots, \mu_4$ be real numbers with $\mu_0 + \ldots + \mu_4 = 2$ and let $x_{\alpha\beta} = \mu_\alpha + \mu_\beta$. Therefore,

$$\sum_{\alpha\beta} x_{\alpha\beta} = 8,$$

and

$$\frac{1}{N}\text{Prop}(Y) = -16 + \frac{1}{2}\sum x_{\alpha\beta}\left\{\Big(\sum_{\{\alpha\beta\}\neq\{\gamma\delta\}} x_{\gamma\delta}D_{\alpha\beta} \cdot D_{\gamma\delta}\Big) + 2x_{\alpha\beta}\right\}.$$

$$(3.14)$$

If we take, for example, $\alpha = 0$, $\beta = 2$, the expression in the brackets $\{\cdot\}$ in the formula above gives

$$x_{34} + x_{13} + x_{14} + 2x_{02} = 4.$$

TABLE 3.1

n_{01}	n_{02}	n_{03}	n_{04}	n_{12}	n_{13}	n_{14}	n_{23}	n_{24}	n_{34}
5	5	5	5	5	5	5	5	5	5
8	8	8	8	4	4	4	4	4	4
9	9	9	3	9	9	3	9	3	3
6	6	6	4	6	6	4	6	4	4
12	12	6	6	6	4	4	4	4	3
12	12	12	4	6	6	3	6	3	3
15	15	15	5	5	5	3	5	3	3
24	24	24	8	4	4	3	4	3	3

In this case, we see that

$$\frac{1}{N}\mathrm{Prop}(Y) = -16 + \frac{1}{2}\sum_{\alpha\beta} 4x_{\alpha\beta} = -16 + 16 = 0.$$

A geometrically interesting situation occurs in the case $x_{\alpha\beta} = 1 - 1/b_{\alpha\beta}$ for a positive integer $b_{\alpha\beta}$, already understood in Definition 3.1 of §3.1. Hence, we have the restriction, which we call INT,

$$n_{\alpha\beta} = \left(1 - (\mu_\alpha + \mu_\beta)\right)^{-1} \in \mathbb{Z}_{\geq 1} \qquad \text{INT.} \qquad (3.15)$$

Up to permutation, the eight solutions of INT with $0 < \mu_\alpha < 1$ and $\sum \mu_\alpha = 2$ are displayed in Table 3.1.

We refer to Chapters 5 and 6 for a discussion of the existence of the covering Y in these cases. The example $\mu_\alpha = 2/5$, and hence $b_{\alpha\beta} = 5$, gives rise to a Kummer covering Y of X (see Appendix B).

3.4 SIGNATURE

In the preceding discussion, the numerical invariants $c_1^2(X)$ and $c_2(X)$ of an algebraic surface X played a fundamental role. The Euler number $e(X) = c_2(X)$ is a topological invariant. It turns out that $c_1^2(X)$ is an invariant of the underlying *oriented* four-dimensional manifold. To see this, we first recall the definition of the signature of a compact oriented four-dimensional manifold X, which we assume to be differentiable. Consider the homology group $H_2(X, \mathbb{Z})$. We have a bilinear symmetric pairing

$$H_2(X, \mathbb{Z}) \times H_2(X, \mathbb{Z}) \longrightarrow \mathbb{Z}$$

given by the intersection number

$$(\alpha, \beta) \longrightarrow \alpha \cdot \beta.$$

The number $\alpha \cdot \beta$ vanishes if α or β is a torsion element of the group $H_2(X, \mathbb{Z})$. The pairing is therefore defined on $\mathcal{L} = H_2(X, \mathbb{Z})/H_2(X, \mathbb{Z})_{\text{tors}}$, which is a lattice of rank $b_2(X)$, where $b_2(X)$ is the second Betti number of X. Thus, we have an integral quadratic form

$$\mathcal{L} \times \mathcal{L} \longrightarrow \mathbb{Z},$$

which is unimodular by Poincaré duality. Over the reals, the form can be diagonalized and on the diagonal it has a number of positive entries, say b_2^+, and a number of negative entries, say b_2^-. We have

$$b_2^+ + b_2^- = b_2(X).$$

By definition, the signature of X is

$$\text{sign}(X) := b_2^+ - b_2^-.$$

The signature depends only on the real quadratic form and can be defined in the same way for a differentiable oriented $4k$-dimensional manifold X. According to the signature theorem ([55], §8), $\text{sign}(X)$ can be expressed in terms of Pontryagin characteristic classes depending only on the real tangent bundle of X. The Pontryagin classes can be expressed in terms of Chern classes when X is a complex manifold. For real dimension 4, this gives the formula

$$\text{sign}(X) = \frac{1}{3}(c_1^2(X) - 2c_2(X)). \tag{3.16}$$

Previously it had been conjectured that $\text{sign}(X) \leq 1$ for an algebraic surface. However, the ball quotients of §3.3, Theorem 3.1, give examples of surfaces with arbitrarily large positive signatures. Indeed, for a ball quotient,

$$\text{sign}(X) = \frac{c_2(X)}{3} = \frac{e(X)}{3}. \tag{3.17}$$

Let $Y \to X$ be a good covering. The formulas (3.9) for $c_2(Y)/N$ and (3.10) for $c_1^2(Y)/N$, together with the adjunction formula (3.11), give

$$3\text{sign}(Y)/N = 3\text{sign}(X) - 2\sum_i x_i D_i^2 + \sum_i x_i^2 D_i^2$$

$$= 3\text{sign}(X) - \sum_i (1 - \frac{1}{b_i^2}) D_i^2. \tag{3.18}$$

Recall that $x_i = 1 - \frac{1}{b_i}$. The fact that this formula for the signature involves only the numbers D_i^2, but not $D_i \cdot D_j$ for $i \neq j$, is related to special properties of the signature expressed by the Atiyah-Bott-Singer fixed point theorem [56].

In the case of the complete quadrilateral, the surface X is $\mathbb{P}_2$ with four points blown up. There are ten divisors on X given by $D_{\alpha\beta}$ with $D_{\alpha\beta}^2 = -1$. We renumber them as D_i, $(i = 1, \ldots, 10)$. We have

$$3\mathrm{sign}(Y)/N = 1 - \sum_{i=1}^{10} \frac{1}{b_i^2}. \tag{3.19}$$

In the case where $\mathrm{Prop}(Y) = 0$ (see the above list),

$$e(Y)/N = 1 - \sum_{i=1}^{10} \frac{1}{b_i^2}. \tag{3.20}$$

If, for example, we have $b_i = 5$ for all i, then

$$c_2(Y)/N = e(Y)/N = \frac{3}{5}.$$

3.5 BLOWING UP POINTS

Let X be a complex surface. Blowing up a point $p \in X$ is a local process, so, using local coordinates, we need only consider the case of blowing up the origin $0 = (0, 0) \in \mathbb{C}^2$. The result is independent of the local coordinates used. Let (z_1, z_2) be the standard coordinates of $\mathbb{C}^2$, and let $(u_1 : u_2)$ be projective coordinates in $\mathbb{P}_1$. Let U be a neighborhood of $(0, 0) \in \mathbb{C}^2$. The blow-up at $(0, 0) \in \mathbb{C}^2$ is given by the subset V of $U \times \mathbb{P}_1$, with coordinates $(z_1, z_2) \times [u_1 : u_2]$ satisfying $z_1 u_2 = z_2 u_1$. For each choice of z_1, z_2, the corresponding points in the blow-up consist of the lines passing through (z_1, z_2) and the origin. If (z_1, z_2) is not the origin, there is a unique such line. If (z_1, z_2) is the origin, the line is not unique; rather, we have all the lines through the origin, corresponding to a copy of $\mathbb{P}_1$. In this way, by blowing up at $p \in X$, we construct a new surface Y and a map

$$\pi : Y \to X.$$

We call $\pi^{-1}(p) = E$ the exceptional divisor. It is a copy of $\mathbb{P}_1$, and

$$\pi : Y \setminus E \to X \setminus \{p\}$$

is biholomorphic. Locally, Y may be described in terms of affine coordinates. Let U be a coordinate neighborhood of p with local coordinates z_1, z_2 centered at p. Then, $W = \pi^{-1}(U)$ is covered by two coordinate charts W_1 and W_2. On W_1 we have coordinates (u, v), and π is given by

$$z_1 = uv, \qquad z_2 = v.$$

On W_2 we have coordinates (u', v'), and π is given by

$$z_1 = u', \qquad z_2 = u'v'.$$

We see that u is an affine coordinate for $(u_1 : u_2)$ on $u_2 \neq 0$ and that v' is an affine coordinate for $(u_1 : u_2)$ on $u_1 \neq 0$. We identify $W_1 \setminus \{u = 0\}$ with $W_2 \setminus \{v' = 0\}$ by the equations $u = \frac{1}{v'}$, $v = u'v'$. The exceptional divisor is given in W_1 by $v = 0$ and in W_2 by $u' = 0$. If L is a smooth irreducible curve through p, then we can suppose that it is given by $z_1 = 0$ in a neighborhood of p, where z_1, z_2 are local coordinates. After blow-up, it is transformed into another curve $\widetilde{L}$, called its proper transform. The curve $\widetilde{L}$ is given by the closure in Y of $\pi^{-1}(L \setminus \{p\})$. Working in W_1, the exceptional divisor E has the equation $v = 0$, and $\widetilde{L}$ has the local equation $u = 0$. Therefore, we have

$$\pi^*L = \widetilde{L} + E, \qquad \widetilde{L} \cdot E = 1, \tag{3.21}$$

and, more generally, the divisor class group of Y is given by

$$\mathrm{Div}(Y) = \pi^*(\mathrm{Div}(X)) \oplus \mathbb{Z}E, \tag{3.22}$$

the direct sum indicating orthogonality with respect to the intersection pairing. For divisors D_1 and D_2 on X, we have $\pi^*D_1 \cdot \pi^*D_2 = D_1 \cdot D_2$. These facts hold because, topologically, homology classes on X can be given by cycles not passing through p. Hence,

$$(\widetilde{L} + E) \cdot E = 0, \tag{3.23}$$

and so we see that E has self-intersection -1:

$$E \cdot E = -1. \tag{3.24}$$

On an algebraic surface, a smooth curve of genus 0, that is, a rational curve, which also has self-intersection -1, is called an *exceptional curve*.

Let us consider the blowing-up process topologically. We have

$$H_2(Y, \mathbb{Z}) = H_2(X, \mathbb{Z}) \oplus \mathbb{Z}[E],$$

and, as the Betti numbers are the ranks of the homology groups, this implies that

$$b_2(Y) = b_2(X) + 1. \tag{3.25}$$

All other homology groups, and hence Betti numbers, are unchanged.

Now consider two real oriented compact differential 4-manifolds M and N, and let $p \in M$ and $q \in N$. Let $U(p)$ be an open neighborhood of p and $U(q)$ be an open neighborhood of q. Suppose that both $\overline{U(p)}$ and $\overline{U(q)}$ are compact balls with respect to a coordinate system. Then, the boundaries of both $M \setminus U(p)$ and $N \setminus U(q)$ are diffeomorphic to a 3-sphere S^3. By definition, the manifold $M\#N$ is that obtained by identifying these boundaries by an orientation *reversing* diffeomorphism. The blow-up of a complex surface X at a point p gives a complex surface Y topologically isomorphic to $X\#\overline{\mathbb{P}_2(\mathbb{C})}$. Here, the orientation on $\mathbb{P}_2(\mathbb{C})$ is reversed and, in particular, a line in $\mathbb{P}_2(\mathbb{C})$ with self-intersection $+1$ becomes a line with self-intersection -1. Only one of these lines is realized complex analytically: the blown-up point. This topological interpretation explains why (3.24) and (3.25) hold.

For manifolds of real dimension 2, the blow-up of the origin over a compact disk around the origin in $\mathbb{R}^2$ yields a Möbius strip, the origin becoming a copy of $\mathbb{P}_1(\mathbb{R})$, the "backbone" of the Möbius strip, and the proper transforms of the lines through the origin intersecting the backbone at the point corresponding to their slope. The Möbius strip is fibered by intervals over $\mathbb{P}_1(\mathbb{R}) \simeq S^1$, just as the blown-up four-dimensional disk is fibered by 2-disks over $\mathbb{P}_1(\mathbb{C}) \simeq S^2$.

Returning to the complex case, we have, for the canonical class,

$$K_Y = \pi^* K_X + E. \tag{3.26}$$

If we write a meromorphic 2-form ω in local coordinates as

$$\omega = a(z_1, z_2) dz_1 \wedge dz_2,$$

then, using the affine coordinates introduced earlier,

$$\pi^* \omega = a(uv, v) v \, du \wedge dv,$$

which proves (3.26). It also follows that

$$K_Y \cdot E = -1. \tag{3.27}$$

We can check this using the adjunction formula, which gives

$$-K_Y \cdot E - E \cdot E = e(E) = e(\mathbb{P}_1) = 2. \tag{3.28}$$

We can also calculate the self-intersection of $\widetilde{L}$ in Y, namely,

$$\widetilde{L} \cdot \widetilde{L} = (\pi^*L - E)^2 = L \cdot L + E \cdot E = L^2 - 1, \qquad (3.29)$$

since E and π^*L are orthogonal with respect to the intersection product. Hence, the self-intersection of L decreases by 1 when we pass to the proper transform. Using (3.22), (3.24), and (3.26), we also have

$$K_Y^2 = c_1^2(Y) = K_X^2 - 1 = c_1^2(X) - 1, \qquad (3.30)$$

whereas, for the Euler numbers,

$$c_2(Y) = c_2(X) + 1, \qquad (3.31)$$

since blowing up removes a point and inserts a copy of S^2. This also follows from the behavior of the Betti numbers in (3.25). Therefore,

$$3c_2(Y) - c_1^2(Y) = 3c_2(X) - c_1^2(X) + 4. \qquad (3.32)$$

Chapter Four

Algebraic Surfaces and the Miyaoka-Yau Inequality

In this chapter we continue our discussion of complex algebraic surfaces, concentrating on the Miyaoka-Yau inequality and the rough classification of surfaces (ROC, for short). Every complex algebraic surface is birationally equivalent to a smooth surface containing no exceptional curves. The latter is called a *minimal surface*. Two related birational invariants, the plurigenus and the Kodaira dimension, play a crucial role in our distinguishing between complex surfaces. Along with the self-intersection number of the canonical divisor, they are part of the ROC of minimal surfaces in Theorem ROC, §4.1. There, we follow closely the terminology used in [58]. The reader may also refer to the classification using the Kodaira dimension given in [12], p. 188, Table 10. This table also includes classes of surfaces that are not algebraic. Our account of ROC is brief, as this is standard material found in most books on complex surfaces. The Miyaoka-Yau inequality, which we discuss in §§4.2 and 4.3, states that for all surfaces of general type, $c_1^2 \leq 3c_2$. To prove this inequality, it suffices to establish it for minimal surfaces. By work of Hirzebruch (1956) [54] and Yau (1978) [144], the two-dimensional free ball quotients are precisely the surfaces of general type with $c_1^2 = 3c_2$.

4.1 ROUGH CLASSIFICATION OF ALGEBRAIC SURFACES

Unless otherwise stated, we denote by X a complex, connected, compact smooth algebraic surface. Recall from Chapter 3, §3.5, that a smooth curve E on X is called exceptional if it has genus 0 and self-intersection number $E^2 = -1$. By Castelnuovo's criterion ([44], Chapter 3, Theorem 9), there is a smooth surface X_0 and a point $p_0 \in X_0$ such that the blow-up of X_0 at p_0 is X. In other words, the exceptional curve E can be blown down to a point p_0, and X_0 is called the *contraction* or *blow-down* of X along E, or simply the contraction or blow-down of E. By a result of Grauert [48], this blowing down theorem is also true for general, not necessarily compact, complex surfaces. Exceptional curves on a surface X may intersect. For example, if we take $\mathbb{P}_1 \times \mathbb{P}_1$ and a point p, then the blow-up at p contains three exceptional curves, namely, the curve E coming from p and the proper transforms of the horizontal and vertical lines $\mathbb{P}_1$ through p. Moreover, if we blow down the two latter proper transforms, we get $\mathbb{P}_2$. Topologically, this corresponds to the relation (see Chapter 3, §3.5)

$$(S^2 \times S^2)\#\overline{\mathbb{P}_2} \simeq \mathbb{P}_2\#\overline{\mathbb{P}_2}\#\overline{\mathbb{P}_2}.$$

Two algebraic surfaces are called *birationally equivalent* if their fields of meromorphic functions are isomorphic. By an important result, two surfaces are birationally equivalent if and only if an iteration of blow-ups of one surface is isomorphic to an iteration of blow-ups of the other surface ([44], Chapter 3, Theorem 12). A surface is called *minimal* if it does not contain any exceptional curves. Every surface X is therefore birationally equivalent to a minimal surface. Indeed, if X is not minimal, we can blow down an exceptional curve to obtain another surface X_1. New exceptional curves may be created by some of the curves passing through the image, p, of the exceptional curve. If X_1 is not minimal we can, once again, blow down an exceptional curve. This process must stop because, upon blowing down, the second Betti number decreases by 1 (see Chapter 3, (3.25)). This blowing down process and the resulting minimal surface may not be unique because exceptional curves may intersect. For example, from $\mathbb{P}_2$ with two points blown up, we can obtain both the minimal surfaces, $\mathbb{P}_2$ and $\mathbb{P}_1 \times \mathbb{P}_1$, by blow-downs.

A holomorphic differential form ω on an algebraic surface X is given locally by

$$\omega = a(z_1, z_2)dz_1 \wedge dz_2,$$

where the coefficient $a(z_1, z_2)$ is holomorphic. As we saw in Chapter 3, §3.2, any such nonzero form determines a nonnegative canonical divisor K; that is, K is a linear combination $\sum a_i C_i$ of irreducible curves with all $a_i \geq 0$. For any positive integer m, consider forms given locally by

$$a(z_1, z_2)(dz_1 \wedge dz_2)^{\otimes^m}.$$

The forms $(dz_1 \wedge dz_2)^{\otimes^m}$ transform under a complex change of coordinates by multiplication by the mth power of the Jacobian of the coordinate change. These forms are called holomorphic 2-forms of weight m. Their totality forms a vector space. We define the *canonical ring* of X to be the graded ring

$$R(X) = \oplus_{m \geq 0} H^0(X, K^{\otimes^m}).$$

Definition 4.1 *The dimension of the vector space of holomorphic 2-forms of weight m on X is called the mth plurigenus $P_m(X) = \dim H^0(X, K^{\otimes^m})$.*

The plurigenus is finite for all m. This follows from theorems of Kodaira on the finiteness of the dimension of spaces of sections of holomorphic bundles

[55]. For $m = 1$, the plurigenus is called the geometric genus and is denoted by g_2 (or p_g). The vector space of holomorphic 2-forms of weight m, and, in particular, the mth plurigenus, is a birational invariant: If we blow up a point p on X, a form can be lifted to the blown-up surface. Conversely, a holomorphic form on the blown-up surface determines a holomorphic form on $X \setminus \{p\}$ given by a holomorphic coefficient $a(z_1, z_2)$ on $U \setminus \{p\}$, where U is a coordinate neighborhood of p. Then the coefficient $a(z_1, z_2)$ can be extended holomorphically to p. For algebraic surfaces, and even for complex analytic surfaces, this follows from Hartog's extension theorem ([66], Theorem 2.3.2).

A nonzero holomorphic 2-form of weight m determines a nonnegative divisor $D = \sum a_i C_i$, $a_i \geq 0$, which is linearly equivalent to mK and called a *multicanonical divisor* of weight m.

A related birational invariant is the *Kodaira dimension*, which associates to an algebraic surface a measure of the growth of the plurigenera.

Definition 4.2 *The Kodaira dimension* $\kappa(X)$ *of an algebraic surface* X *is defined as follows. If* $P_m(X) = 0$ *for all* $m \geq 0$, *set* $\kappa(X) = -\infty$. *Otherwise, let*

$$\kappa(X) = \min\{k \in \mathbb{Z}, k \geq 0 : P_m(X)/m^k \text{ is bounded as a function of } m\}.$$

If $P_m(X) \neq 0$ for some m, then, as $P_{nm}(X) \geq P_m(X)$, it follows that $\kappa(X) \geq 0$. Equivalent definitions of $\kappa(X)$ are the transcendence degree of the canonical ring $R(X)$, minus 1; and for m large enough, the dimension of the image of X under the *pluricanonical map*, which is determined by the linear system of divisors associated to $K^{\otimes m}$ and maps to projective space of dimension $P_m - 1$. It follows that $\kappa(X) = -\infty, 0, 1, 2$, since $\dim(X) = 2$. By a result of Enriques ([44], Chapter 10, Corollary 6), we know that a surface X is rational or ruled if and only if $\kappa(X) = -\infty$. In these cases, the minimal models are known but may not be unique. A surface is ruled if it is birationally equivalent to $C \times \mathbb{P}_1$, where C is a smooth curve. If $C = \mathbb{P}_1$, the surface is rational and, as remarked earlier, is birational to $\mathbb{P}_2$. If $\kappa(X) = 2$, the canonical ring $R(X)$ is finitely generated, and the projective variety $\text{Proj}(R(X))$ is called the *canonical model* of X ([10], Chapter 9).

Remark. The condition $\kappa(X) \geq 0$ is equivalent to the assumption that there exists a nonnegative multicanonical divisor D of weight $m \geq 1$. The analogue for algebraic curves, that is, Riemann surfaces, is the assumption that $g \geq 1$. Here are some consequences of this assumption:

a) Let $D \geq 0$ be a multicanonical divisor for some $m \geq 1$. Every exceptional curve E on the surface X is contained in D. Indeed, we have $K \cdot E = -1$. Therefore $mK \cdot E = -m < 0$ and $D \cdot E < 0$. If E were not contained in D, this intersection, $D \cdot E$, would be nonnegative, which is a contradiction.

b) Suppose two distinct exceptional curves E_1 and E_2 intersect on X. Write the multicanonical divisor as

$$D = \ldots + n_1 E_1 + \ldots + n_2 E_2 + \ldots,$$

with n_1, $n_2 > 0$; then,

$$D \cdot E_1 \geq -n_1 + n_2 E_1 \cdot E_2$$

and

$$D \cdot E_2 \geq -n_2 + n_1 E_1 \cdot E_2$$

with $E_1 \cdot E_2 > 0$, so that either $D \cdot E_1$ or $D \cdot E_2$ is nonnegative, which is a contradiction (see a)). Hence, no two exceptional curves in X intersect.

c) Therefore, there are only finitely many disjoint exceptional curves. Blowing all of them down gives a surface with $\kappa(X) \neq -\infty$; the exceptional curves are again disjoint and can be blown down. This uniquely determined process must terminate because the second Betti number decreases each time. By a result of Enriques and Kodaira, the *minimal surface* obtained in this way is, up to isomorphism, unique in its birational equivalence class ([44], Chapter 3, Theorem 19).

Minimal algebraic surfaces are roughly classified in Theorem ROC below. This result and some of the explanations can be found in [58], where, in turn, the basic reference is [87]. In contrast to [58], we indicate the Kodaira dimension for each class and we use the terminology "properly elliptic" for the "honestly elliptic" surfaces with Kodaira dimension 1 ([12], p. 188, Table 10).

Theorem 4.1 (Theorem ROC) *An algebraic surface without exceptional curves belongs to exactly one of the following classes:*

1) *the projective plane* $\mathbb{P}_2$ *($K^2 = 9$, $\kappa(X) = -\infty$),*
2) *the surfaces* Σ_n, $n \geq 0, n \neq 1$ *($K^2 = 8$, $\kappa(X) = -\infty$),*
3) *the geometrically ruled surfaces (algebraic* $\mathbb{P}_1$*-bundles) over a nonsingular algebraic curve of genus at least 1 ($K^2 = 8(1 - g) \leq 0$, $\kappa(X) = -\infty$),*
4) *the algebraic K3-surfaces ($K^2 = 0$, $\kappa(X) = 0$),*
5) *the abelian surfaces (two-dimensional algebraic tori) ($K^2 = 0$, $\kappa(X) = 0$),*
6) *the honestly elliptic surfaces without exceptional curves; there are two cases,*

    (a) *the Enriques surface (quotient of a K3-surface by a fixed point free involution) and the hyperelliptic surface (quotient of an abelian surface by a finite group acting freely) ($K^2 = 0, \kappa(X) = 0$),*

(b) *the properly elliptic surfaces without exceptional curves* ($K^2 = 0$, $\kappa(X) = 1$),

7) *the minimal surfaces of general type* ($K^2 > 0$, $\kappa(X) = 2$).

We now explain the terminology used in this theorem. The surfaces Σ_n are algebraic fibrations over $\mathbb{P}_1$, with $\mathbb{P}_1$ as fiber and $\mathbb{C}^*$ as structural group. The group $\mathbb{C}^*$ fixes 0 and ∞ on $\mathbb{P}_1$. Therefore Σ_n has two disjoint sections. They are rational curves with self-intersection numbers n and $-n$, respectively. These well-known ruled surfaces were studied by F. Hirzebruch in his thesis [53]. The surface Σ_0 is $\mathbb{P}_1 \times \mathbb{P}_1$. The surface Σ_1 is $\mathbb{P}_2$ with a point blown up; therefore it is excluded in the theorem. The surface Σ_{n+1} is obtained from Σ_n by first blowing up a point on the intersection of some fiber with the section of self-intersection number $-n$, and, next, by blowing down the proper transform of the fiber. The surfaces of the classes 1) and 2) are precisely the rational surfaces without exceptional curves. A K3 surface is, by definition, a surface X with first Betti number 0 and with a trivial canonical class ($K = 0$). Every K3 surface has Euler number 24. An elliptic surface is a surface that admits at least one elliptic fibering, that is, a holomorphic map onto a nonsingular curve such that all but a finite number of fibers are nonsingular elliptic curves. Some of the surfaces in the classes 1)–5) and some of their blow-ups admit elliptic fiberings. By an honestly elliptic surface we mean an elliptic surface not birationally equivalent to a surface belonging to any of the classes 1)–5). The properly elliptic surfaces are the honestly elliptic surfaces of Kodaira dimension 1.

The surfaces in the classes 1), 2), and 3) have $\kappa(X) = -\infty$, and we can always find birationally equivalent surfaces in which two exceptional curves meet. If the surface is in class 3), we blow up a point. The exceptional curve coming from the blown-up point meets the proper transform of the fiber through the point. This proper transform is also exceptional. The surfaces in classes 4), 5), 6), and 7) satisfy $\kappa(X) \geq 0$. In view of Theorem ROC, we could define a surface of general type as one that satisfies $\kappa(X) \geq 0$ and whose minimal model satisfies $K^2 > 0$. From Theorem ROC, we also see that a surface, minimal or not, with $K^2 > 9$ is of general type.

As mentioned above, we have an alternative definition of general type. Consider the vector space of holomorphic 2-forms of weight m on X and let $\omega_1, \omega_2, \ldots, \omega_{P_m(X)}$ be a basis, where we suppose $P_m(X) \geq 1$. Then $(\omega_1 : \omega_2 : \ldots : \omega_{P_m(X)})$ is by definition the m-canonical map

$$\phi_m : X \to \mathbb{P}_{P_m(X)-1}(\mathbb{C}).$$

This mapping is defined except at those points of X where all holomorphic 2-forms of weight m vanish; for example, the points on exceptional curves. An equivalent definition of the Kodaira dimension is the maximum dimension of the images $\phi_m(X)$, $m \geq 1$, which we set equal to $-\infty$ if $P_m(X) = 0$. Again, we see that the Kodaira dimension is at most 2. A surface of general type is, by definition, a surface for which there is at least one m such that ϕ_m provides a birational equivalence onto a, possibly singular, algebraic surface. In fact, according to Bombieri [16], for a minimal surface of general type, the map ϕ_m, $m \geq 5$, is defined everywhere and has the following properties. Let C be a smooth rational curve with self-intersection number -2. Then $K \cdot C = 0$ by the adjunction formula. Therefore, for every nonnegative m-canonical divisor D, either C is contained in D or it is disjoint from D. It follows that either ϕ_m is not defined on C or it is constant on C. Therefore, when $m \geq 5$, the result of Bombieri implies that ϕ_m maps C to a point. Moreover, let A be the union of all smooth rational curves with self-intersection -2 and let $A_1, \ldots, A_l$ be the connected components of the set A. Then, for $m \geq 5$, ϕ_m provides a biholomorphic map from $X \setminus A$ onto its image, and each set A_i is mapped onto a normal singular point of the image, with different A_i's carried to different points. We shall come back to this in the next section. We also remark that for any irreducible curve C that is not contained in A, we have $K \cdot C > 0$.

4.2 THE MIYAOKA-YAU INEQUALITY, I

In Chapter 3, we introduced the quantity

$$\text{Prop}(X) = 3c_2(X) - c_1^2(X) = 3e(X) - K_X^2.$$

If we blow up a point on X, then $\text{Prop}(X)$ increases by 4. For the cases 1), 2), 4), and 5) of Theorem ROC, the Euler number is 3, 4, 24, and 0, respectively, and $K^2 = 9, 8, 0$, and 0. Therefore, $\text{Prop}(X) \geq 0$ for all rational surfaces and for all blow-ups of surfaces in classes 4) and 5). In class 6), the Euler number is greater than or equal to 0, and $K^2 = 0$. Therefore, $\text{Prop}(X) \geq 0$ for all surfaces with minimal model in class 6). In class 3), the Euler number equals $4(1 - g)$, where g, is the genus of the base curve, whereas $K^2 = 8(1 - g)$. Therefore, for $g \geq 2$ we expect surfaces with $\text{Prop}(X) < 0$. An example is given by the Cartesian product $X = C \times \mathbb{P}_1$, where C is a smooth curve of genus $g \geq 2$. Here,

$$K \simeq -2C \times \{\text{point}\} \cup \{\text{point}\} \times (2g - 2)\mathbb{P}_1,$$

so that $K^2 = -8(g - 1)$, $e(X) = -4(g - 1)$ and $\text{Prop}(X) = -4(g - 1) < 0$. The result of Miyaoka-Yau shows that $\text{Prop}(X) \geq 0$ for all surfaces of general type, minimal or not. Hence, surfaces with $\text{Prop}(X) < 0$ have minimal models in class 3) with a base curve of genus at least 2. We shall now concentrate on

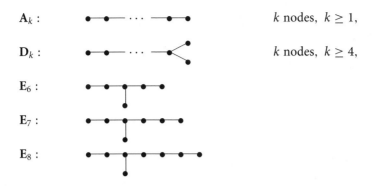

$\mathbf{A}_k$: k nodes, $k \geq 1$,

$\mathbf{D}_k$: k nodes, $k \geq 4$,

$\mathbf{E}_6$:

$\mathbf{E}_7$:

$\mathbf{E}_8$:

Figure 4.1 Dynkin diagrams for the connected components of A

the Miyaoka-Yau inequality for surfaces of general type, giving a rough outline of some of the ideas involved and providing references for the details.

Miyaoka and Yau independently investigated the inequality named after them. Miyaoka used methods from algebraic geometry, for example, Hodge's index theorem, as well ideas of F.A. Bogomolov; see [139], [140]. Yau used differential geometry. Some important ingredients are due independently to Aubin. We do not say anything about the algebro-geometric approach, but in §4.3, as well as in Chapters 5 and 6, we do discuss proofs of Miyaoka-Yau-type inequalities using methods from differential geometry.

We nonetheless make a few remarks about Miyaoka's work [104]. He proved that $\mathrm{Prop}(X) \geq 0$ for all surfaces X of general type. As mentioned before, it suffices to demonstrate the inequality for minimal surfaces, because $\mathrm{Prop}(X)$ decreases by 4 upon blowing down an exceptional curve. Miyaoka refines the inequality for $\mathrm{Prop}(X)$ using the union A of all smooth rational curves with self-intersection -2 on X; see [57]. The intersection matrix $\|A_i \cdot A_j\|_{i,j}$ of the connected components A_i of A is negative definite (see, for example, [10], p. 124, Claim 9.1.2, [12], Chapter III). These connected components are therefore associated to the well-known Dynkin diagrams shown in Figure 4.1 below. There, each node corresponds to a smooth rational curve with self-intersection -2 and each edge to a transversal intersection at 1 point. No other intersections occur ([23], [12], Chapter 1, Lemma 2.12).

Let $\mathcal{E}$ be such a configuration. If we blow down all the curves to a point, we get a rational double point that is the quotient singularity of $\mathbb{C}^2$ modulo a finite subgroup G of $\mathrm{SU}(2)$. In Chapters 1 and 2, we introduced the finite subgroups of $\mathrm{SO}(3)$, namely, D_v (dihedral group of order $2v$), the automorphism group T of the tetrahedron, and W of the cube, and I of the icosahedron. They can be lifted to $\mathrm{SU}(2)$ via the projection $\mathrm{SU}(2) \rightarrow \mathrm{SO}(3)$, and give the subgroups D'_v,

T', W', and I'. These have orders $4v$, 24, 48, and 120, respectively. The group G corresponding to the configuration $\mathbf{A}_k$ is the cyclic group of order $k + 1$. It is the group D'_{k-2} for $\mathbf{D}_k$. Furthermore G equals T', W', and I' for $\mathbf{E}_6$, $\mathbf{E}_7$, and $\mathbf{E}_8$ ([35]).

To each configuration $\mathcal{E}$, we associate a number $m(\mathcal{E})$, given by

$$m(\mathcal{E}) = 3(k + 1) - \frac{3}{|G|},$$

where k is the number of nodes and G is the corresponding group. The stronger inequality may be described as follows.

Let X be a minimal surface of general type with $A = \cup A_i$, where the A_i are the connected components of A and, therefore, configurations of the type $\mathbf{A}_k$, $\mathbf{D}_k$, $\mathbf{E}_6$, $\mathbf{E}_7$, $\mathbf{E}_8$. Then,

$$\mathrm{Prop}(X) \geq \sum m(A_i),$$

which is positive if A is not empty.

We shall prove this stronger inequality in a special case. Namely, let Y be an algebraic surface on which a finite group H operates by automorphisms. Suppose that H has only finitely many fixed points and that the isotropy group at each fixed point is isomorphic to one of the groups G above. We resolve the singularities of the orbit space Y/H by blowing up each of them into one of the configurations defined by the Dynkin diagrams in Figure 1. The surface Y/H so resolved is an algebraic surface X. If X is of general type, then Y is too. Let p run through the singularities of Y/H and let G_p be the isotropy group at one of the points of Y representing p. Then, by standard Euler number arguments,

$$e\left(Y/H \setminus \cup\{p\}\right) = \frac{1}{|H|}\left(e(Y) - \sum_p \frac{|H|}{|G_p|}\right).$$

Let $\mathcal{E}_p$ be the configuration of 2-spheres to which the singularity p of Y/H is blown up. Then, the Euler number, $e(\mathcal{E}_p)$, equals the number of nodes in $\mathcal{E}_p$ plus 1. Therefore,

$$\frac{3e(Y)}{|H|} = 3e(X) - \sum_p m(\mathcal{E}_p).$$

Since there are multicanonical divisors not intersecting the $\mathcal{E}_p$, we also have

$$\frac{c_1^2(Y)}{|H|} = c_1^2(X).$$

Then, we have

$$\text{Prop}(X) = \frac{\text{Prop}(Y)}{|H|} + \sum_p m(\mathcal{E}_p)$$

and

$$\text{Prop}(X) \geq \sum_p m(\mathcal{E}_p),$$

because the Miyaoka-Yau inequality is valid for Y.

In fact, the above inequality is true for surfaces of general type, minimal or not, if the $\mathcal{E}_p$ are among the configurations above and occur disjointly on the surface. This statement is somewhat more general than the strengthened inequality involving the whole of A.

Before describing Yau's methods of proof, let us note the following. We want to characterize those surfaces of general type X that are ball quotients by the condition $\text{Prop}(X) = 0$. If X contains a smooth rational curve C, then it cannot be a ball quotient. Indeed, the curve C would lift to the ball and would give rational curves. This is impossible because the coordinate functions z_1, and z_2 of the ball would be constant on these compact curves. If C were a smooth elliptic curve on X, it would lift to coverings of C, which are C, $\mathbb{C}^*$, or $\mathbb{C}$. By Liouville's theorem, the bounded coordinate functions would, again, be constant in the cases $\mathbb{C}^*$ and $\mathbb{C}$. Therefore, X cannot be a ball quotient. If smooth rational or elliptic curves occur, the Miyaoka-Yau inequality cannot be sharp.

4.3 THE MIYAOKA-YAU INEQUALITY, II

This short overview of the proof of the Miyaoka-Yau inequality, using tools from complex analysis and differential geometry, is based on the presentation in [20]. We shall indicate some additional references as needed. By the discussion in §4.1 for a compact connected smooth algebraic surface X, the canonical divisor class is ample if and only if X is of general type and is minimal, and the union A of all smooth rational curves on X with self-intersection -2 is empty. We shall make all these assumptions on X until otherwise stated at the end of this section. For $m \geq 5$, the pluricanonical map ϕ_m of §4.1 maps X biholomorphically onto its image in a projective space.

The hard part of the proof of the Miyaoka-Yau inequality is the existence of a so-called Kähler-Einstein metric on X, defined below. To conclude the proof, one uses a result of Guggenheimer for such metrics that follows from straightforward tensor computations.

Following the conventions of differential geometry, our local coordinates will now have upper indices. Let Y be a complex manifold of dimension n. We reproduce in this paragraph, for convenience, some of the material of Chapter 1, §§1.3 and 1.4, to which we refer you for more details. A metric on Y is given locally by a tensor

$$ds^2 = \sum_{k,\ell=1}^{n} g_{k\bar{\ell}} dz^k d\bar{z}^\ell = \sum_{k,\ell=1}^{n} g_{k\bar{\ell}} dz^k \otimes d\bar{z}^\ell,$$

where the $g_{k\bar{\ell}}$ are smooth functions. Assume the metric is Hermitian, that is, at every point the $n \times n$ matrix with (k, ℓ)-entry, $g_{k\bar{\ell}}$ is positive definite and Hermitian ($g_{k\bar{\ell}} = \overline{g_{\ell\bar{k}}}$). To such a metric we associate the $(1, 1)$-form,

$$\omega = \frac{\sqrt{-1}}{2\pi} \sum_{k,\ell=1}^{n} g_{k\bar{\ell}} dz^k \wedge d\bar{z}^\ell.$$

As $g_{k\bar{\ell}} = \overline{g_{\ell\bar{k}}}$, we have $\omega = \bar{\omega}$, so that ω is a $(1, 1)$-form that is a real 2-form on Y. Such a form is called a real $(1, 1)$-form. The metric is defined to be Kähler when the form ω is closed, that is, $d\omega = 0$. The form ω is then called the Kähler form and is a representative of a class $[\omega]$ in $H^2(Y, \mathbb{R})$. Assume from now on that Y is a compact Kähler manifold. Recall that a Kähler metric can be written locally as

$$g_{k\bar{\ell}} = \partial^2 K / \partial z^k \partial \bar{z}^\ell$$

for some smooth real-valued function K, called the Kähler potential.

As we saw in Chapter 1, §1.3, projective space $\mathbb{P}_N$ carries a Kähler metric, called the Fubini-Study metric. In terms of homogeneous coordinates

$$[z_0 : z_1 : \ldots : z_N]$$

of $\mathbb{P}_N$, it is given in the coordinate patch $z_i \neq 0$ by the Hermitian matrix

$$FS_{k\bar{\ell}} = \frac{\left(1 + \sum_k |w_k|^2\right) \delta_{k\ell} - \overline{w}_k w_\ell}{\left(1 + \sum_k |w_k|^2\right)^2} = \partial^2 \log\left(1 + \sum_k |w_k|^2\right) / \partial w_k \partial \overline{w}_\ell,$$

where $w_k = z_k/z_i, k = 1, \ldots, N, k \neq i$, are the affine coordinates. As, by our assumptions, the surface X is holomorphically immersed in a projective space via the pluricanonical map, the Fubini-Study metric induces a Kähler metric on X, since the associated $(1, 1)$-form on X is the restriction of the $(1, 1)$-form of the Fubini-Study metric (we discuss this further, below).

A real $(1, 1)$-form ω on a complex manifold Y is called *positive* if, for every holomorphic tangent vector v on Y, we have $-\sqrt{-1}\omega(v, \bar{v}) \geq 0$, and *positive definite* if, whenever $v \neq 0$, this inequality is strict. This enables us to define inequalities between real $(1, 1)$-forms. The associated form of a Hermitian metric is a positive $(1, 1)$-form since, on the holomorphic tangent space at $y \in Y$, we have, in local coordinates, $g(\lambda \partial / \partial z^k, \mu \partial / \partial z^\ell)|_y = \lambda \overline{\mu} g_{k\overline{\ell}}(y)$.

On the other hand, we can associate to the metric its Ricci form, given locally by

$$\gamma_1 = \left(\sqrt{-1}/2\pi\right) \sum_{k,\ell} R_{k\overline{\ell}} dz^k \wedge d\overline{z}^\ell,$$

where

$$R_{k\overline{\ell}} = -\partial^2 \log \det \left(g_{k\overline{\ell}}\right) / \partial z^k \partial \overline{z}^\ell.$$

This is a closed real $(1, 1)$-form. It is also called the Chern form as the first Chern class $[c_1(Y)] \in H^2(Y, \mathbb{Z})$ of Y is represented in $H^2(Y, \mathbb{R})$ by $[\gamma_1]$.

A Kähler metric is called Kähler-Einstein when its Ricci form γ_1 is proportional to the Kähler form ω. Therefore, there is a (real) constant λ such that $\gamma_1 = \lambda \omega$. In other words, $R_{k\overline{\ell}} = \lambda g_{k\overline{\ell}}$, that is,

$$-\partial^2 \log \det \left(g_{k\overline{\ell}}\right) / \partial z^k \partial \overline{z}^\ell = \lambda g_{k\overline{\ell}}.$$

These partial differential equations are called the Einstein equations. They imply the strong restriction that the first Chern class is definite (when $\lambda < 0$ or when $\lambda > 0$) or identically zero (when $\lambda = 0$), in the sense that the associated symmetric bilinear form given by γ_1 has these properties. The Fubini-Study metric on projective space $\mathbb{P}_N$ is Kähler-Einstein with factor $\lambda = N + 1$. As already remarked, the restriction of the Fubini-Study metric defines a Kähler metric on any complex submanifold of projective space. However, the restriction of a Kähler-Einstein metric to a submanifold is not necessarily Kähler-Einstein. A smooth real function ρ on a chart U of an atlas of a complex manifold with local coordinates z^k is defined to be strictly plurisubharmonic (on U) if the matrix $(\partial^2 \rho / \partial z^k \partial \overline{z}^\ell)(p)$ is positive definite for all $p \in U$. Therefore, the local Kähler potentials of a Kähler metric give a system of locally strictly plurisubharmonic functions on Y. Let Z be a complex submanifold of Y. We can choose charts on Y and on Z such that the submanifold Z is locally described by the vanishing of some of the local coordinates of Y. The matrix of partial derivatives $(\partial^2 / \partial z^k \partial \overline{z}^\ell)$ on Z is a principal minor of the corresponding matrix on Y, restricted to Z. Therefore, any system of locally plurisubharmonic functions on Y induces, by restriction to Z, a system of locally plurisubharmonic functions on Z. The local Kähler potentials on Y therefore induce, by restriction, local Kähler potentials on Z, and hence define a Kähler metric on Z. This amounts to restricting the Kähler form on Y to Z, thereby obtaining a Kähler form of a Kähler metric on Z.

We can also introduce the concept of the Ricci form of a metric using volumes (see also Chapter 1, §1.4). A smooth volume form Ω on a complex manifold Y of dimension n is a positive smooth (n, n)-form. In terms of local coordinates z_α^i, $i = 1, \ldots, n$, on a chart U_α of an atlas of Y, a volume form is given locally by

$$\Omega = h_\alpha \prod_{i=1}^{n} \sqrt{-1}\, dz_\alpha^i \wedge d\bar{z}_\alpha^{\,i},$$

with $h_\alpha > 0$ and $h_\alpha = |\det(\partial z_\beta^i / \partial z_\alpha^j)|^2 h_\beta$ on $U_\alpha \cap U_\beta$. The *Ricci form* $\mathrm{Ric}\, \Omega$ of Ω is the real $(1, 1)$-form given, locally by

$$\mathrm{Ric}\, \Omega = -\sqrt{-1}\, \partial\bar{\partial} \log \Omega$$

$$=: -\sqrt{-1} \sum_{k,\ell=1,\ldots,n} \left(\partial_k \partial_{\bar{\ell}} \log h_\alpha\right) dz_\alpha^k \wedge d\bar{z}_\alpha^\ell,$$

where $\partial_k = \partial/\partial z_\alpha^k$ and $\partial_{\bar{\ell}} = \partial/\partial \bar{z}_\alpha^\ell$. Notice that we do not divide by 2π in the definition of the Ricci form. In different references, conventions about sign and normalization factors vary. We have adopted those closest to [84], our main reference in Chapter 6. If ω is the $(1, 1)$-form associated to a Hermitian metric $g_{k\bar{\ell}}$ on Y as in Chapter 1, §1.4 (so there is no 2π), then the wedge product ω^n of ω with itself n times satisfies

$$\frac{\omega^n}{n!} = \det(g_{k\bar{\ell}}) \prod_{i=1}^{n} \sqrt{-1}\, dz^i \wedge d\bar{z}^i,$$

where the positivity of $\det(g_{k\bar{\ell}})$ follows from the fact that the matrix $(g_{k\bar{\ell}})$ is a positive definite Hermitian matrix, and so has for all its eigenvalues positive real numbers. Therefore, $\omega^n/n!$ is a volume form on Y. The Ricci form of this volume form is 2π times the Ricci form γ_1 of the metric, so that

$$\mathrm{Ric}\left(\frac{\omega^n}{n!}\right) = 2\pi \gamma_1.$$

For example, on $\mathbb{P}_N$, the volume form $\Psi = \psi^N/N!$, where ψ is the $(1, 1)$-form associated to the Fubini-Study metric, satisfies $\mathrm{Ric}(\Psi) = (N + 1)\psi$. Therefore, a Kähler metric is Kähler-Einstein when its $(1, 1)$-form satisfies

$$\mathrm{Ric}\left(\frac{\omega^n}{n!}\right) = \lambda \omega,$$

for some constant λ. As noted above, the Ricci forms of any two Kähler metrics ω, ω' on Y define the same cohomology class in $H^2(Y, \mathbb{R})$. Therefore, there exists a smooth real-valued function f on Y, unique up to an additive constant, such that

$$\mathrm{Ric}(\omega'^n/n!) = \mathrm{Ric}(\omega^n/n!) - \sqrt{-1}\partial\bar{\partial} f.$$

On the other hand, as Y is orientable, we have $\omega'^n = F\omega^n$ for some smooth positive function F on Y. Therefore,

$$\sqrt{-1}\partial\bar{\partial} \log F - \sqrt{-1}\partial\bar{\partial} f = \sqrt{-1}\partial\bar{\partial} \left(\log F - f\right) = 0.$$

We deduce that $\log F - f = \log A$, where $A > 0$ is constant on Y, so that

$$\omega'^n = Ae^f \omega^n.$$

As ω and ω' define the same cohomology class in $H^2(Y, \mathbb{R})$,

$$\int_Y \omega'^n = A \int_Y e^f \omega^n,$$

which determines the constant A.

Let's consider again the complex algebraic surface X, which is compact, smooth, connected, and holomorphically embedded in a projective space $\mathbb{P}_N$ by a pluricanonical mapping ϕ_m, $m \geq 5$. Therefore, the canonical class K_X of X is ample. Let $g_{k\bar{\ell}}$ be the metric induced by restriction to X of the Fubini-Study metric of $\mathbb{P}_N$, and let ω be the Kähler $(1, 1)$-form (so there is a 2π). The Fubini-Study metric on $\mathbb{P}_N$ has for integral cohomology class the first Chern class of the holomorphic line bundle $\mathcal{O}_{\mathbb{P}_N}(1)$ on $\mathbb{P}_N$ that is associated to the canonical divisor class on $\mathbb{P}_N$ (see Chapter 1, §1.4). The bundle on X associated to the pluricanonical divisor $K^{\otimes m}$ is the pullback $\phi_m^*(\mathcal{O}_{\mathbb{P}_N}(1))$. Therefore, the $(1, 1)$-form $\frac{1}{m}\omega$ has cohomology class $[\frac{1}{m}\omega] = -c_1(X) = c_1(K) \in H^2(X, \mathbb{Q})$. In particular, we have $c_1(X) < 0$. Incidentally, the Kodaira embedding theorem [86] characterizes projective complex manifolds inside the class of compact Kähler manifolds. It states that a compact complex manifold Y is projective if and only if Y admits a Kähler class that is rational, that is, belongs to $H^2(Y, \mathbb{Q}) \subset H^2(Y, \mathbb{R})$. The proof of the "only if" part of that result uses the fact that if the Kähler class $[\omega]$ of a Kähler metric on Y is rational, then some multiple of $[\omega]$ is integral and therefore equals the Chern class of a holomorphic line bundle, which is then shown, using nontrivial arguments, to be ample. If a connected smooth compact complex surface Z carries a negative Kähler-Einstein metric (i.e. $\lambda < 0$), the negative first Chern class, $-c_1(Z)$, of Z can be represented by the 2-form of a Kähler metric. Moreover, as $-c_1(Z) = c_1(K_Z)$, these arguments of Kodaira show that the canonical line bundle K_Z of Z is ample. Therefore, some multiple of K_Z defines a holomorphic embedding of Z in projective space. The result we shall outline, proved by T. Aubin [7] and, independently, by S. T. Yau [144], states that a connected compact complex surface Z admits a Kähler-Einstein metric with $\lambda < 0$ if and only if $c_1(Z) < 0$. The Kähler-Einstein metric is uniquely determined by Z and the complex structure on Z. By the remarks at the beginning of this paragraph, their results apply to X (with the given assumptions).

Our goal, then, is to find an Kähler-Einstein metric on X. To do this, we will find a Kähler-Einstein metric $\widetilde{g}_{k\bar{\ell}}$ whose associated $(1, 1)$-form is in the same cohomology class as the associated $(1, 1)$-form of $g_{k\bar{\ell}}$. Therefore, $\widetilde{g}_{k\bar{\ell}}$ will be a solution of the Einstein equation.

Let's see how the solution of the Einstein equation for a Kähler metric can be reduced to a so-called Monge-Ampere equation (we follow [2], §3). Given a Kähler metric $g = (g_{k\bar{\ell}})$ with Kähler form ω, we look for a Kähler-Einstein metric $\widetilde{g} = (\widetilde{g}_{k\bar{\ell}})$ with Kähler form $\widetilde{\omega}$ such that $\widetilde{g}_{k\bar{\ell}} = g_{k\bar{\ell}} + \partial_k \partial_{\bar{\ell}} u$ for a smooth function u on X. Here, again, we let $\partial_k = \partial/\partial z^k$ and $\partial_{\bar{\ell}} = \partial/\partial \bar{z}^\ell$, and $\partial = \sum_{k=1,2} \partial_k dz^k$, $\bar{\partial} = \sum_{k=1,2} \partial_{\bar{k}} d\bar{z}^k$, and $\partial\bar{\partial} = \sum_{k,\ell} (\partial_k \partial_{\bar{\ell}}) dz^k \wedge d\bar{z}^\ell$, are thought of as operators applied to functions. Expressing $g_{k\bar{\ell}}$ in terms of its Kähler potential K, and using the definition of $R_{k\bar{\ell}}$, we see that there is locally a smooth function F such that $R_{k\bar{\ell}} - \lambda g_{k\bar{\ell}} = \partial_k \partial_{\bar{\ell}} F$, namely,

$$F = -\log \det(g_{k\bar{\ell}}) - \lambda K.$$

Now, if $\widetilde{g}_{k\bar{\ell}} = g_{k\bar{\ell}} + \partial_k \partial_{\bar{\ell}} u$ for a smooth function u on X, the Einstein equation for $\widetilde{g}$,

$$\widetilde{R}_{k\bar{\ell}} = -\partial^2 \log \det(\widetilde{g}_{k\bar{\ell}})/\partial z^k \partial \bar{z}^\ell = \lambda \widetilde{g}_{k\bar{\ell}},$$

is equivalent to the scalar complex Monge-Ampere equation

$$\left(\det(g_{k\bar{\ell}})\right)^{-1} \det(g_{k\bar{\ell}} + \partial_k \partial_{\bar{\ell}} u) = e^{-\lambda u + F}.$$

To see this, take $\partial_k \partial_{\bar{\ell}}$ of the logarithm of both sides, to obtain

$$-\widetilde{R}_{k\bar{\ell}} + R_{k\bar{\ell}} = -\lambda \partial_k \partial_{\bar{\ell}} u + \partial_k \partial_{\bar{\ell}} F =$$

$$-\lambda \left(\widetilde{g}_{k\bar{\ell}} - g_{k\bar{\ell}}\right) + \left(R_{k\bar{\ell}} - \lambda g_{k\bar{\ell}}\right) = -\lambda \widetilde{g}_{k\bar{\ell}} + R_{k\bar{\ell}},$$

which implies $\widetilde{R}_{k\bar{\ell}} = \lambda \widetilde{g}_{k\bar{\ell}}$ as required. Note that, when $u = 0$, we have $R_{k\bar{\ell}} = \lambda g_{k\bar{\ell}}$, so that we can take $F = 0$ and the Monge-Ampere equation still makes sense. One checks directly that $\widetilde{\omega}$ is in the same class as ω in $H^2(X, \mathbb{R})$. Therefore, there is a Kähler-Einstein metric in the same class as ω if and only if there is a solution u of the scalar complex Monge-Ampere equation with $\widetilde{g}_{k\bar{\ell}} = g_{k\bar{\ell}} + \partial_k \partial_{\bar{\ell}} u$ positive definite.

Using the Monge-Ampere equation, it was proved by T. Aubin [7] and, independently, by S. T. Yau [144] that a compact complex surface X admits a

Kähler-Einstein metric with $\lambda < 0$ if and only if $c_1 < 0$. The Kähler-Einstein metric is uniquely determined by X and the complex structure on X. The proof is a deep analytic theorem using "a priori" estimates. In particular, one shows that there is a solution u and a positive constant c such that

$$c\omega < \tilde{\omega} = \omega + \sqrt{-1}\partial\bar{\partial}u < c^{-1}\omega.$$

If X carries a negative Kähler-Einstein metric ($\lambda < 0$), the negative first Chern class can be represented by the 2-form of a Kähler metric, and, by Kodaira, this is equivalent to saying that the m-pluricanonical map gives a projective embedding of X, for m sufficiently large. In other words, the canonical divisor K is ample. Recall that the negative first Chern class is the cohomology class of the canonical bundle. By the discussion of §4.1, for algebraic surfaces X, the divisor class is ample if and only if X is of general type and minimal, and the union A of all smooth rational curves on X with self-intersection -2 is empty.

The Chern classes c_1 and c_2 of a smooth compact complex surface X, with a Hermitian metric g, can be represented by Chern differential forms γ_1 and γ_2 in real de Rham cohomology. The forms γ_k are real (k, k) forms, defined explicitly in terms of the metric g, such that the Chern numbers c_2, c_1^2 of the surface are given by the integrals

$$c_1^2(X) = \int_X \gamma_1^2 := \int_X \gamma_1 \wedge \gamma_1, \qquad c_2(X) = \int_X \gamma_2.$$

The forms γ_1 and γ_2 are defined as follows. To the Hermitian metric g we associate a connection ∇ on the tangent space with curvature R. In a local coordinate system $z = (z^1, z^2)$, with the associated tangent space basis $\{\partial/\partial z^1, \partial/\partial z^2\}$, we can describe ∇ by a 2×2 matrix of $(1, 0)$-forms $\vartheta = (\vartheta_j{}^i)$, and R by a 2×2 matrix $\Theta = (\Theta_j{}^i)$ of $(1, 1)$-forms. Namely,

$$\vartheta = (\partial g)g^{-1}, \qquad \Theta = \bar{\partial}\vartheta,$$

where $g = (g_{j\bar{k}})$ is the coefficient matrix of the metric (see [67], 1.3, 1.5). The forms γ_k are the real (k, k) forms

$$\gamma_1 = (\sqrt{-1}/2\pi)\left(\Theta_1{}^1 + \Theta_2{}^2\right) = (\sqrt{-1}/2\pi)\mathrm{Trace}(\Theta)$$

and

$$\gamma_2 = (-1/8\pi^2)\left(\Theta_1{}^1 \wedge \Theta_2{}^2 - \Theta_2{}^1 \wedge \Theta_1{}^2\right) = -(1/8\pi^2)\det(\Theta),$$

where we require the metric to be normalized so that, if ω is the Kähler form of g, we have

$$\int_X \omega^2 := \int_X \omega \wedge \omega = 1.$$

The form γ_1 is given by the Ricci form above, and γ_2 is the so-called Gauss–Bonnet form.

Guggenheimer [50] observed in 1952 that, for a Kähler-Einstein metric, we have $0 \leq \gamma_1^2 \leq 3\gamma_2$, and the equality $\gamma_1^2 = 3\gamma_2$ holds exactly when the holomorphic sectional curvature is constant. The proof is not difficult. First, we modify a given choice of local coordinates by a linear transformation in such a way that the metric at the origin 0 is the standard metric $g_{k\bar{\ell}}(0) = \delta_{k\bar{\ell}}$. From the Einstein condition $R_{k\bar{\ell}} = \lambda g_{k\bar{\ell}}$, one deduces that $R_{k\bar{\ell}}(0) = \lambda \delta_{k\bar{\ell}}$. Using simple tensor coefficient manipulations and the definition of the Chern forms, we deduce that there are real numbers a, b, with $a + b = \lambda$, and a complex number c such that

$$\gamma_1^2 = (-2/4\pi^2)(a+b)^2 dz^1 \wedge d\bar{z}^1 \wedge dz^2 \wedge d\bar{z}^2,$$

$$\gamma_2 = (-1/4\pi^2)(a^2 + 2b^2 + 4c\bar{c})\, dz^1 \wedge d\bar{z}^1 \wedge dz^2 \wedge d\bar{z}^2.$$

From this, we deduce the claim of the theorem at the origin: we check directly that, for $a, b \in \mathbb{R}$ and $c \in \mathbb{C}$, we have

$$2(a+b)^2 \leq 3(a^2 + 2b^2 + 4c\bar{c}).$$

Indeed, a rearrangement of the above inequality gives

$$0 \leq (a - 2b)^2 + 12c\bar{c},$$

so that we have equality exactly when $a = 2b$ (so that $\lambda = 3b = 3a/2$) and $c = 0$.[1] In this case, from its definition, the holomorphic sectional curvature has the value $S = 2\lambda/3$ at the origin. Now, any point of the surface can be viewed as the origin of a local coordinate system. Therefore, the above inequalities hold everywhere. When the equality $\gamma_1^2 = 3\gamma_2$ holds everywhere, it follows (see [85], II: IX, 7.5, p. 168) that the holomorphic sectional curvature on the total bundle of the complex tangent directions is constant (and equal $2\lambda/3$).

By integrating $3\gamma_2 - \gamma_1^2$ over X, the existence of a Kähler-Einstein metric therefore implies the Miyaoka-Yau inequality

$$3c_2(X) \geq c_1^2(X) \geq 0,$$

[1] In [11], it is erroneously stated that equality occurs for $a = b$, $c = 0$.

and the special form of the integrand $3\gamma_2 - \gamma_1^2$ shows that we have equality $3c_2(X) = c_1^2(X)$ if and only if the holomorphic sectional curvature on X is constant, that is, the Gaussian curvature in the complex directions is constant throughout X. Recall that a Hermitian metric determines a Riemannian metric, with curvature as defined in Chapter 2, §2.3, when restricted to a locally defined two-dimensional real differentiable submanifold. Its curvature at $p \in X$ depends only on p and on a two-dimensional submanifold of the tangent space of X at p. This subspace is called complex if it is a one-dimensional subspace of the complex tangent space. For a Kähler metric, ω^k is, up to a positive factor, the volume element for a complex k-dimensional submanifold. For a negative Kähler-Einstein metric, the complex curvature is negative everywhere.

If the algebraic surface X has an ample canonical class and if the Miyaoka-Yau inequality is sharp, that is, $3c_2(X) = c_1^2(X)$, then the universal cover $\tilde{X}$ of X is the ball B_2 given by

$$|z^1|^2 + |z^2|^2 < 1.$$

For the proof, note that a simply-connected complex surface carrying a Kähler-Einstein metric with constant negative holomorphic curvature is isometric to the ball.

The metric on the ball is given by

$$ds^2 = \sum_{k,\ell} \frac{\partial^2 \log \mathcal{K}}{\partial z^k \partial \bar{z}^\ell} dz^k \otimes d\bar{z}^\ell,$$

where

$$\mathcal{K} = (1 - |z^1|^2 - |z^2|^2)^{-1}.$$

It is invariant under the group of all biholomorphic maps of the ball, which is isomorphic to PU(2, 1) by classical theorems. Every biholomorphic map is projective and can be extended to $\mathbb{P}_2$. Therefore, the fundamental group Γ of X is represented as a discrete subgroup of PU(2, 1), and X is isomorphic to $\Gamma \backslash B_2$. Observe that the isotropy group of PU(2, 1) at the origin is U(2). It follows that PU(2, 1) acts transitively on all complex directions anywhere in the ball, and, therefore, the holomorphic curvature is constant.

If an algebraic surface has a Kähler metric with negative holomorphic curvature, then all smooth curves on the surface have a negative Euler number, so there are no smooth rational and no smooth elliptic curves on the surface. Previously, we made this observation for ball quotients.

We saw in §4.2 that smooth rational curves A_i with self-intersection -2 on X contribute a positive number $m(A_i)$ to the lower bound for Prop(X).

Smooth elliptic curves on X also contribute a positive number to this lower bound. The following theorem, formulated in [57], gives an improved Miyaoka-Yau inequality. It is a special case of results of R. Kobayashi [81], [82], Y. Miyaoka [105], F. Sakai [122], and S. T. Yau [143].

Theorem 4.2 *Let X be a minimal surface of general type and A the union of smooth curves of self-intersection -2 in X with connected components A_i (so of type $\mathbf{A}_k$, $\mathbf{D}_k$, $\mathbf{E}_6$, $\mathbf{E}_7$, $\mathbf{E}_8$). Suppose that $C_1, \ldots, C_r$ are smooth elliptic curves on X that are disjoint from each other and also disjoint from A. Then,*

$$\mathrm{Prop}(X) \geq \sum_i m(A_i) + \sum_{j=1}^r (-C_j \cdot C_j).$$

By the adjunction formula, $-C_j \cdot C_j = K_X \cdot C_j$, and this number is positive, as X is assumed to be of general type (see §4.1). If equality holds in the above inequality, then there exists a discrete subgroup Γ of $\mathrm{PU}(1, 2)$ such that

$$X' \setminus \cup C_j = \Gamma \backslash B_2,$$

where X' is the singular surface obtained from X by blowing down the A_i. The blow-downs of the A_i give quotient singularities corresponding to non-trivial subgroups of Γ that are isotropy groups of points in the ball. These isotropy groups are all subgroups of $\mathrm{SU}(2)$. The group Γ has r "cusps," that is, fixed points on the boundary of B_2. If we compactify $\Gamma \backslash B_2$ by adding a point for each cusp, the compactification $\overline{\Gamma \backslash B_2}$ has r singular points that are resolved by the r elliptic curves C_j. The surface X is the smooth model, that is, the minimal desingularization, of $\overline{\Gamma \backslash B_2}$.

For more details on why the cusps are resolved by elliptic curves, see the discussion of isotropy subgroups of Γ for points on the boundary of B_2 in Chapter 6. The above result can be proven either using techniques from algebraic geometry that generalize those mentioned in §4.2, as in [122], or using methods from differential geometry generalizing those of this section, as in [81], [82]. These latter techniques are similar to those discussed in Chapter 6. In Chapter 5, we also treat in more detail the case where there are elliptic curves on X.

   J. C. Hemperly [52] was the first to study the singularities of surfaces $\overline{\Gamma \backslash B_2}$. For Γ the Picard modular group, an extensive study of these surfaces was carried out in a series of publications by R.-P. Holzapfel [60], [61], [62], [63], [64]. We summarize briefly some of the results of [52], which show the appropriate Miyaoka-Yau equality for a non-compact ball quotient. Let Y_0 be a non-compact quotient of the ball B_2 by a discrete subgroup Γ of $\mathrm{PU}(1, 2)$

acting freely that is of finite covolume with respect to the ball metric on B_2. As remarked already, the open surface Y_0 can be compactified by adding a finite number of points P, called "cusps," which correspond to fixed points of Γ on the boundary of B_2, and the resulting surface is normal (any resolution of its singularities is a smooth complex projective variety). Then, by replacing Γ by a subgroup of finite index if necessary, we can resolve each singularity P by an elliptic curve C_P that has negative self-intersection in the resulting smooth surface Y. Let C be the union of all such elliptic curves. From [52], §5.7, we have

$$(K_Y + C)^2 = K_Y^2 - C \cdot C = 3c_2(Y \setminus C) = 3e(Y) > 0.$$

(This can be interpreted as saying that the "logarithmic" Chern classes of (Y, C) are proportional to those of the projective plane: see Chapter 6.) As in the discussion of the Miyaoka-Yau inequality in this section, and of its generalizations due to S. Y. Cheng and S. T. Yau [27] and R. Kobayashi [84], discussed in Chapters 5 and 6, Hemperly proved the above equality by integrating the Chern forms of a suitable metric on Y. On Y_0, this metric is the hyperbolic metric induced by the ball metric on B_2, which is invariant under $PU(1,2)$, and hence under Γ. To extend it to Y, he used Grauert's criterion [48] to observe that each cusp P is obtained from a disk bundle on the elliptic curve C_P by blowing down the zero section. For each cusp P there is a countable basis of punctured neighborhoods $U_{P,n,0}$, that give punctured disk bundles over the compactifying elliptic curve C_P. On each of these bundles, $U_{P,n,0}$, one can construct a positive definite Hermitian metric that can be extended to the full disk bundle $U_{P,n}$, with the curve C_P as zero section. The resulting metrics on the "annulus bundles"

$$V_{P,n} := U_{P,n} \setminus \overline{U}_{P,n+1}$$

can be glued to each other with the help of a partition of unity; and, together with the hyperbolic metric of the open ball quotient surface Y_0, they give a C^∞-metric on the smooth compact surface Y. One then defines the Chern differential forms γ_1^2 and γ_2 associated with this metric (just as we did earlier in this section), whose integrals can be evaluated. Hemperly showed that the limit values (for $n \to \infty$) of the integrals

$$\int_{V_{P,n}} \gamma =: \gamma[V_{P,n}]$$

of these Chern forms over the annulus bundle $V_{P,n}$ exist and are given by

$$\lim_{n\to\infty} \gamma_1^2[V_{P,n}] = C_P^2, \qquad \lim_{n\to\infty} \gamma_2[V_{P,n}] = 0.$$

Since on the complement of the disk bundle we still have the hyperbolic metric with $\gamma_1^2 = 3\gamma_2$, it follows at once (with $U_n := \cup_P U_{P,n}$ and $V_n := U_n \setminus \overline{U}_{n+1}$) that

$$K_Y^2 = \lim_{n \to \infty} \gamma_1^2[Y \setminus U_{n+1}] = \lim_{n \to \infty} \left(\gamma_1^2[Y \setminus U_n] + \gamma_1^2[V_n] \right)$$

$$= \lim_{n \to \infty} 3\gamma_2[Y \setminus U_n] + \lim_{n \to \infty} \gamma_1^2[V_n] = 3e(Y) + \sum_P c_P^2,$$

as required.

Chapter Five

Line Arrangements in $\mathbb{P}_2(\mathbb{C})$ and Their Finite Covers

As we discussed in Chapter 4, by work of Hirzebruch [54] and Yau [144] we know that the two-dimensional free ball quotients are precisely those surfaces of general type with $c_1^2 = 3c_2$. In this chapter, we derive necessary conditions for weighted line arrangements in $\mathbb{P}_2$ to admit finite covers that are ball quotients by finding solutions to $c_1^2 = 3c_2$ for such finite covers. More precisely, these are finite covers of the blow-ups of the line arrangements at their singular intersection points. We list all the known weighted line arrangements satisfying these necessary conditions. The weights will be determined by certain diophantine conditions. The line arrangements we study are listed in §5.3. They all satisfy the condition that the number of intersection points on each line of the arrangement is $(k + 3)/3$, where k is the number of lines of the arrangement. It is not known if the list we give here is the complete list of line arrangements with this property. For each line arrangement, we give the table of possibilities for the weights in the course of this chapter. In this chapter, we do not address the question of whether the corresponding finite covers of the blown-up line arrangements actually exist. We shall discuss this existence question in Chapter 6.

We begin with a few general remarks. Consider an arrangement of $k \geq 2$ lines in $\mathbb{P}_2$. Any two lines of the arrangement have a nonempty intersection. If exactly two lines of the arrangement pass through a point, it is called a *regular point*. If more than two lines of the arrangement pass through a point, it is called a *singular point*. An arrangement is said to be in *general position* if it has only regular points. The number of such points is then $t_2 = k(k-1)/2$. For an arbitrary arrangement, let t_r denote the number of points through which pass r lines. Then, $k(k-1)/2$ is the total number of intersection points counted with multiplicities. We have

$$\frac{k(k-1)}{2} = \sum_{r \geq 2} \frac{r(r-1)}{2} t_r. \tag{5.1}$$

To understand this formula, notice that, on deforming the lines passing through a point of intersection of r lines, we can form $r(r-1)/2$ regular points. We assume from now on that $t_k = 0$, that is, the arrangement does not form a pencil.

As §§5.1 and 5.2 consist of general arguments for line arrangements, it is instructive to first consider some examples. For an integer $q \geq 2$, consider the arrangement in $\mathbb{P}_2$ defined by the equations

$$(x_2^q - x_1^q)(x_2^q - x_0^q)(x_0^q - x_1^q) = 0, \tag{5.2}$$

where x_0, x_1, x_2 are homogeneous coordinates on $\mathbb{P}_2$. This is called the Ceva(q) *arrangement*. It has $k = 3q$ lines. The case $q = 1$ would give a pencil of three lines, that is, $t_3 = 1$, which we have excluded. For $q \neq 3$, the arrangement has $t_3 = q^2$, $t_q = 3$, and $t_r = 0$ otherwise. When $q = 3$, it has nine lines and twelve triple points. When $q = 2$, we have the arrangement Ceva(2), which is equivalent to the complete quadrilateral; see Chapter 3, Figure 3.1. These arrangements are named after Giovanni Ceva (1647–1734), who found a criterion for three lines, one through each vertex of a triangle, to pass through a common point. Menelaos of Alexandria, circa 100 A.D., formulated the dual problem. Ceva's criterion can be formulated as follows. The equation $x_0 x_1 x_2 = 0$ represents a triangle. Let $c_0 = x_1 / x_2$ parametrize the lines through $x_1 = x_2 = 0$, let $c_1 = x_2 / x_0$ parametrize the lines through $x_0 = x_2 = 0$, and let $c_2 = x_0 / x_1$ parametrize the lines through $x_0 = x_1 = 0$. Then, taking a line l_i with parameter c_i from each family, these three lines intersect at a point if and only if $c_0 c_1 c_2 = 1$. The three lines $x_i = 0$ together with the three lines l_i with $c_i = 1$ form the complete quadrilateral arrangement. In the Ceva(q) arrangement, each l_i is replaced by q lines parametrized by the qth roots of unity, and, by using the above criterion, it is easy to see which groups of three lines intersect. However, the three lines $x_i = 0$ do not belong to Ceva(q).

By adding the three lines $x_i = 0$ to the Ceva(q) arrangement, we can form the *extended* Ceva(q) *arrangement*, denoted by $\overline{\text{Ceva}}(q)$. It has the equation

$$x_0 x_1 x_2 (x_2^q - x_1^q)(x_2^q - x_0^q)(x_0^q - x_1^q) = 0, \tag{5.3}$$

and, for $q \neq 1$, it has $k = 3q + 3$, $t_2 = 3q$, $t_3 = q^2$, $t_{q+2} = 3$, and $t_r = 0$ otherwise. When $q = 1$ we have the arrangement $\overline{\text{Ceva}}(1)$, which is isomorphic to the complete quadrilateral, and so has $k = 6$, $t_2 = 3$, $t_3 = 4$, with $t_r = 0$ otherwise.

Another arrangement, the *icosahedral arrangement*, is obtained as follows from the icosahedron inscribed in the 2-sphere S^2. Project the icosahedron from the origin onto S^2 and identify the antipodes, that is, consider $S^2/\{\pm 1\}$, which is isomorphic to $\mathbb{P}_2(\mathbb{R})$. The twelve vertices of the icosahedron become six points in $\mathbb{P}_2(\mathbb{R})$, which form the five vertices of a regular pentagon together with its center. The fifteen lines connecting pairs of these six points correspond to the 30 edges of the icosahedron. For this arrangement, we therefore have $k = 15$, $t_2 = 15$, $t_5 = 6$, $t_3 = 10$, with $t_r = 0$ otherwise. The ten triple points correspond to the midpoints of the twenty faces of the icosahedron or to the vertices of the dual dodecahedron. The fifteen regular points correspond to the midpoints of the 30 edges of the icosahedron.

It was conjectured by J. J. Sylvester in 1893, and only proved in 1933 by Tibor Gallai, that an arrangement of $k \geq 3$ lines in real projective space that do not all pass through one point ($t_k = 0$), that is, do not form a pencil, necessarily has $t_2 > 0$. Therefore, the arrangements with $t_2 = 0$, for example, Ceva(q) with $q \geq 3$, cannot occur over the reals.

In this chapter, we study finite coverings of $\mathbb{P}_2$ that are branched along line arrangements such as those given in the above examples. However, the discussion of Chapter 3 was based on the assumption that the branching locus consisted of divisors with normal crossings, and this is certainly not the case for the above examples. In order to be able to apply the considerations of Chapter 3, we construct in §5.1 a new algebraic surface by blowing up the multiple intersection points of the arrangement.

5.1 BLOWING UP LINE ARRANGEMENTS

Consider an arrangement of k lines, $L_1, L_2, \ldots, L_k$, in $\mathbb{P}_2$, and recall that we denoted by t_r the number of points lying at the intersection of exactly r lines of the arrangement. Consider a new complex surface $X \to \mathbb{P}_2$ obtained by blowing up all points that are singular ($r \geq 3$). We shall use the index j to signify the singular points p_j that when blown up become the *exceptional divisors* E_j. No two of these exceptional divisors intersect. We reserve the index i for the smooth divisors L_i, whose proper transforms we denote by D_i. On X, the divisors D_i and E_j have normal crossings, and we can therefore apply the considerations of Chapter 3. Let $m_j \geq 1$ be a preassigned ramification along E_j, and $n_i \geq 1$ a preassigned ramification along D_i. Let

$$x_i = 1 - \frac{1}{n_i} \tag{5.4}$$

and

$$z_j = 1 - \frac{1}{m_j}. \tag{5.5}$$

The self-intersection numbers are

$$E_j \cdot E_j = -1 \tag{5.6}$$

and

$$D_i \cdot D_i = 1 - \sigma_i, \tag{5.7}$$

where σ_i is the number of singular points of the arrangement lying on L_i. This follows from Chapter 3 (3.29), and the fact that a line in $\mathbb{P}_2$ has self-intersection 1. We study coverings Y of X of degree N that are good coverings in the sense of Definition 3.1 of Chapter 3 with vanishing Prop. We revise the notation of that definition and express (i), (ii), and (iii) in terms of the D_i's and E_j's. We rewrite, for X as above, Definition 3.1, from Chapter 3, without recopying the local coordinate descriptions

Definition 5.1 *Let Y be a complex surface that is a finite covering*

$$\pi : Y \to X$$

of X—that is, above any point of X there are only a finite number of points of Y—and that is branched along the system of divisors, $\{D_i\}_{i \in I}$, and exceptional divisors, $\{E_j\}_{j \in J}$. Suppose that Y is a good covering in the sense of Definition 3.1, Chapter 3, §3.1. Then, there are integers $N \geq 1$ and $n_i \geq 2$, $m_j \geq 1$, $i \in I$, $j \in J$, such that

(i) *over each point of $D_i \setminus \cup_{i \neq k} D_i \cap D_k \setminus \cup_{i,j} D_i \cap E_j$ there are N/n_i points of Y and over each point of $E_j \setminus \cup_{i,j} D_i \cap E_j$ there are N/m_j points of Y;*

(ii) *over each point of a nonempty intersection $D_i \cap D_k$, $i \neq k$, there are $N/n_i n_k$ points of Y, and over each point of a nonempty intersection $D_i \cap E_j$ there are $N/n_i m_j$ points of Y;*

(iii) *over the points not appearing in (i) and (ii) there are N points of Y. We call N the* degree *of the covering. At any such point of Y, the map π is locally biholomorphic.*

5.2 HÖFER'S FORMULA

Let $Y \to X$ be a good covering of degree N as in Definition 5.1. Using formula (3.12) of Theorem 3.2, Chapter 3, for $(3c_2(Y) - c_1^2(Y))/N$, together with the formulas of §5.1 of this chapter, and the fact, shown in Chapter 3, §3.5, that every blown-up point contributes $+4$ to $3c_2(X) - c_1^2(X)$, it follows that

$$(3c_2(Y) - c_1^2(Y))/N = 4\sum_j 1 + \sum_i x_i(-2\sigma_i) + \sum_j z_j(-4) + \sum_i x_i^2(\sigma_i - 1)$$

$$+ \sum_j z_j^2 + \sum_{i,j} x_i z_j D_i \cdot E_j + \frac{1}{2} \sum_{i \neq k} x_i x_k D_i \cdot D_k. \quad (5.8)$$

Here, we also use $e(D_i) = e(E_j) = 2$. The above equation simplifies to give

$$(3c_2(Y) - c_1^2(Y))/N = \frac{1}{4}\sum_j \left(2z_j - 4 + \sum_i x_i D_i \cdot E_j\right)^2$$

$$-\frac{1}{4}\sum_j \left(\sum_i x_i D_i \cdot E_j\right)^2 + \sum_i x_i^2(\sigma_i - 1) + \frac{1}{2}\sum_{i \neq k} x_i x_k D_i \cdot D_k. \quad (5.9)$$

Setting

$$P_j = 2z_j - 4 + \sum_i x_i D_i \cdot E_j, \quad (5.10)$$

we see that we can write the right-hand side of (5.9), which we call "Prop" for short, and denote by $\mathrm{Prop}(Y)/N$, as

$$\mathrm{Prop}(Y)/N = (3c_2(Y) - c_1^2(Y))/N = \frac{1}{4}\sum_j P_j^2 + \frac{1}{4}\sum_{i,l} R_{il} x_i x_l, \quad (5.11)$$

where the matrix $R = (R_{il})$ is given by $R_{ii} = 3\sigma_i - 4$ and

$$\begin{aligned} R_{il} &= -1 & L_i \cap L_l &= \{\text{singular point}\}, \ i \neq l, \\ R_{il} &= 2 & L_i \cap L_l &= \{\text{regular point}\}, \ i \neq l. \end{aligned}$$

This important formula, (5.11), is due to Höfer [11]. It is interesting that Höfer's formula proves the Miyaoka-Yau inequality in the present situation provided R is positive semi-definite. Applying this to the example of the complete quadrilateral, we can again find the solution already encountered in Chapter 3, §3.3, of Prop $= 0$. The matrix R in this case is

$$\begin{pmatrix} 2 & -1 & -1 & -1 & -1 & 2 \\ -1 & 2 & -1 & -1 & 2 & -1 \\ -1 & -1 & 2 & 2 & -1 & -1 \\ -1 & -1 & 2 & 2 & -1 & -1 \\ -1 & 2 & -1 & -1 & 2 & -1 \\ 2 & -1 & -1 & -1 & -1 & 2 \end{pmatrix}.$$

In the notation of Chapter 3, §3.3, the $D_i, i = 1, \ldots, 6$, are to be replaced by $D_{12}, D_{13}, D_{14}, D_{23}, D_{24}, D_{34}$. The matrix R is positive semi-definite. Its rank is 2: it has a 2×2 positive definite sub-matrix. There is a four-dimensional vector space annihilating this matrix given by $x_{\alpha\beta} = \mu_\alpha + \mu_\beta$, where α and $\beta, \alpha \neq \beta$, as before run through $1, \ldots, 4$. We verify that the P_j are also annihilated if j is replaced by $01, 02, 03,$ and 04, where $\mu_0 + \mu_1 + \ldots + \mu_4 = 2$ and

$$z_{0\beta} = \mu_0 + \mu_\beta, \qquad \beta = 1, \ldots, 4.$$

For example, we have

$$2(\mu_0 + \mu_1) - 4 + \mu_2 + \mu_3 + \mu_2 + \mu_4 + \mu_3 + \mu_4 = 0.$$

Returning to a general line arrangement, we now rewrite Höfer's formula in a way that is important for later applications. With the notation of this section, if we let

$$y_j := z_j - 2 = -1 - \frac{1}{m_j},$$

then from (5.10) we have

$$P_j = 2y_j + \sum_i x_i D_i \cdot E_j, \tag{5.12}$$

and $\mathrm{Prop}(Y)/N$ becomes a quadratic form in the variables x_i and y_j, which we denote by $\mathrm{Prop}(x, y)$. We have

$$\frac{\partial}{\partial y_j} \mathrm{Prop}(x, y) = P_j. \tag{5.13}$$

Moreover,

$$P_j = -\frac{2}{m_j} + r_j - 2 - \sum_{i \sim j} \frac{1}{n_i}, \tag{5.14}$$

where $i \sim j$ means that D_i and E_j intersect, and where r_j is the number of lines passing through the singular point indexed by j.

Lemma 5.1 *We have*

$$\frac{\partial}{\partial x_i} \mathrm{Prop}(x, y) = Q_i,$$

with

$$Q_i = \frac{-2(\sigma_i - 1)}{n_i} + \tau_i - 2 - \sum_{j \sim i} \frac{1}{m_j} - \sum_{l \sim i} \frac{1}{n_l}, \tag{5.15}$$

where τ_i is the number of intersection points on the line L_i. The notation $l \sim i$ in the second summation means that D_i and D_l meet at a regular point (which therefore was not blown up).

Proof. By (5.11) and (5.12), we have

$$\frac{\partial}{\partial x_i} \text{Prop}(x, y) = \frac{1}{2} \left(\sum_{j \sim i} 2y_j + \sum_{j \sim i} \sum_{l \sim j} x_l + \sum_{l} R_{il} x_l \right). \qquad (5.16)$$

We have

$$\sum_{l} R_{il} x_l = (3\sigma_i - 4)x_i - \sum_{j \sim i} \sum_{l \sim j} x_l (1 - \delta_{i,l}) + 2 \sum_{i \sim l} x_l, \qquad (5.17)$$

where $\delta_{i,l}$ is Kronecker's delta function, and so, by (5.16) and (5.17),

$$\frac{\partial}{\partial x_i} \text{Prop}(x, y) = \frac{1}{2} \left(2 \sum_{j \sim i} y_j + (4\sigma_i - 4)x_i + 2 \sum_{l \sim i} x_l \right),$$

which is formula (5.15) of the lemma. It follows from (5.13) and (5.15) that

$$\text{Prop}(x, y) = \frac{1}{2} \sum_{i} x_i Q_i + \frac{1}{2} \sum_{j} y_j P_j. \qquad (5.18)$$

There is a uniform way to interpret the P_j and Q_i:

Definition 5.2 *Let D be a smooth compact curve on a complex surface. We define*

$$\text{prop}(D) = 2D \cdot D - e(D).$$

We can motivate Definition 5.2 as follows. For a smooth compact curve D in a ball quotient B_2/Γ, where the group Γ acts freely and discontinuously (with finite covolume), we have $\text{prop}(D) = 0$ if and only if D lifts to a line in the ball on which the restricted action of Γ is also free and discontinuous. For any smooth compact curve in such a ball quotient, we have $\text{prop}(D) \geq 0$. We remark that, for a line L in $\mathbb{P}_2$, the number $\text{prop}(L)$ vanishes. These facts follow from a theorem of Enoki; see also [11], Anhang B.2, H. If we denote the inverse image of D_i in Y by $\widetilde{D_i}$ and of E_j by $\widetilde{E_j}$, then we have

$$\frac{N}{n_i} Q_i = \text{prop}(\widetilde{D_i}).$$

and

$$\frac{N}{m_j} P_j = \text{prop}(\widetilde{E}_j).$$

Therefore,

$$\text{Prop}(Y) = \frac{1}{2}\left(\sum_i (n_i - 1)\text{prop}(\widetilde{D}_i) - \sum_j (m_j + 1)\text{prop}(\widetilde{E}_j)\right). \quad (5.19)$$

5.3 ARRANGEMENTS ANNIHILATING R AND HAVING EQUAL RAMIFICATION ALONG ALL LINES

The vector that has each of its coordinates equal to 1 is in the kernel of the matrix R defined in §5.2 if and only if the sum of the elements in any row of R equals 0. For $i = 1, \ldots, k$, let τ_i be the number of intersection points on the line L_i, and $t_r^{(i)}$ the number of r-fold intersection points on L_i. Rewriting the condition that the sum of the terms on the ith row of R is 0, we find the equivalent condition that, for each $i = 1, \ldots, k$,

$$3\sigma_i - 4 - \sum_{r\geq 3}(r-1)t_r^{(i)} + 2t_2^{(i)} = 3\sigma_i - 4 + 3t_2^{(i)} - \sum_{r\geq 2}(r-1)t_r^{(i)}$$

$$= 3\tau_i - 4 - (k-1) = 3\tau_i - k - 3 = 0. \quad (5.20)$$

Therefore, in order that (5.20) hold, we must have, for each $i = 1, \ldots, k$,

$$\tau = \tau_i = \frac{k+3}{3}. \quad (5.21)$$

Problem: A problem that remains *unsolved* is that of classifying all line arrangements with the property

$$3\tau = 3\tau_i = k + 3, \quad \text{for all } i = 1, \ldots, k.$$

We now list all known examples of such line arrangements.

Triangular arrangement: If $(x_0 : x_1 : x_2)$ are projective coordinates in $\mathbb{P}_2$, a triangular arrangement is given by the three lines $x_0 = 0$, $x_1 = 0$, $x_2 = 0$. We have $k = 3$ and $\tau = 2$, so clearly $3\tau = k + 3$.

Ceva and extended Ceva: For Ceva(2), and $\overline{\text{Ceva}}(1)$, we have $k = 6$ and $\tau = 3$, so (5.21) is satisfied. Indeed, for all $q \geq 2$, it is straightforward to check that (5.21) holds for the configurations Ceva(q) and $\overline{\text{Ceva}}(q)$.

Hesse arrangement: This arrangement has $k = 12$ lines, with $t_2 = 12$, $t_4 = 9$, and $t_r = 0$ otherwise (by formula (5.1)). It can be described as follows. Consider a smooth cubic curve in $\mathbb{P}_2$. If $(x_0 : x_1 : x_2)$ are homogeneous coordinates for $\mathbb{P}_2$ and if $C = C(x_0, x_1, x_2)$ is the homogeneous equation of the cubic $C = 0$, then the Hessian of C is the form

$$H = H(x_0, x_1, x_2) = \det\left(\frac{\partial^2 C}{\partial x_i \partial x_j}\right).$$

The points of inflection of C are the zeros of the forms C and H. The smooth cubic has the structure of an elliptic curve, and one inflexion point can be taken as the origin O for its group structure [130], Chapter 1, §6.2. The sum of the intersection points of an arbitrary line with the cubic is O. If this line is tangent to the cubic with multiplicity 3, and hence is tangent at an inflexion point, this point corresponds to a point of order 3 on the elliptic curve. All inflexion points arise in this way and there are consequently nine inflexion points on the cubic. This can also be seen from Bezout's Theorem. Moreover, a line connecting two inflexion points of the cubic intersects the cubic at a third inflexion point. Therefore, there are twelve lines connecting the nine inflexion points on C. These are the twelve lines of the Hesse arrangement, the inflexion points themselves being its quadruple points, each of the four lines passing through two other inflexion points. The twelve lines form four triangles whose vertices are the twelve double points. Therefore, on each line $L_i, i = 1, \ldots, 12$, of the Hesse arrangement we have $t_2^{(i)} = 2$ and $t_4^{(i)} = 3$. Therefore $\tau = 5$ for all lines and $3\tau = k + 3 = 15$. The position of the nine inflexion points is projectively uniquely determined, as is the pencil of all cubics passing through them. This pencil can be written as

$$\lambda C + \mu H = 0, \quad (\lambda : \mu) \in \mathbb{P}_2,$$

where H, as above, is the Hessian of C. If $C = x_0^3 + x_1^3 + x_2^3$, then the pencil is

$$\lambda(x_0^3 + x_1^3 + x_2^3) + \mu x_0 x_1 x_2 = 0.$$

The four triangles are the four degenerate cubics in the pencil given by the equation

$$C_{12} = x_0 x_1 x_2 \{27 x_0^3 x_1^3 x_2^3 - (x_0^3 + x_1^3 + x_2^3)^3\} = 0. \tag{5.22}$$

The twelve double points are the vertices of the triangle $x_0 x_1 x_2 = 0$, together with the nine projective points $(x_0 : x_1 : x_2)$ with all x_i third roots of unity. These are exactly the twelve triple points of Ceva(3),

$$C_9 = (x_0^3 - x_1^3)(x_1^3 - x_2^3)(x_0^3 - x_2^3) = 0. \tag{5.23}$$

Extended Hesse arrangement: With notation as above, the extended Hesse arrangement is given by the equation

$$C_{12} \cdot C_9 = 0, \tag{5.24}$$

with $k = 21$, $t_2 = 36$, $t_4 = 9$, $t_5 = 12$, and all other $t_r = 0$. The number of r-fold points on a given line depends on whether the line is in $C_{12} = 0$, when we have $t_2^{(i)} = 3$, $t_4^{(i)} = 3$, $t_5^{(i)} = 2$, or in $C_9 = 0$, in which case we have $t_2^{(i)} = 4$, $t_4^{(i)} = 0$, $t_5^{(i)} = 4$. In both cases, $\tau = 8$ and $3\tau = k + 3$ as required.

Icosahedral arrangement: This arrangement appeared in Chapter 2, §2.7, and at the beginning of this chapter. It has $k = 15$ lines and $\tau = 6$. For each $i = 1, \ldots, k$ we have $t_2^{(i)} = t_3^{(i)} = t_5^{(i)} = 2$. Each of the fifteen lines is fixed by one of the fifteen involutions of the icosahedral group A_5 acting on the projective plane.

Klein configuration: This arrangement has $k = 21$ lines that are fixed lines of order 2 of the simple finite group G_{168} with 168 elements referred to in Chapter 2, §2.7. For this arrangement, we have $t_3 = 28$ and $t_4 = 21$, and all other $t_r = 0$. For each $i = 1, \ldots, k$, we have $t_3^{(i)} = 4$, $t_4^{(i)} = 4$, giving $3\tau = k + 3$.

Valentiner arrangement: Finally, consider the group A_6 of order 360. In 1889, H. Valentiner defined an action of A_6 on the projective plane. He proceeded in an analogous way to Klein's treatment of G_{168}, realizing A_6 as the quotient of a hyperbolic triangle group of signature $(2, 4, 5)$ by a normal congruence subgroup of level 3 and index 360. The group A_6 acts on the corresponding modular curve given by the quotient of the upper half plane by the congruence subgroup. This curve can be embedded in the projective plane as a smooth curve of degree 6, and this enables us to realize an A_6 action on the projective plane. Full details can be found in R. Fricke's *Lehrbuch der Algebra*, Volume 2, Vieweg, Braunschweig 1926. The Valentiner group can be realized explicitly as the subgroup of $GL(3, \mathbb{C})$ given by the complex reflections ([132]):

$$R_1 = \begin{pmatrix} 0 & -\omega^2 & 0 \\ -\omega & 0 & 0 \\ 0 & 0 & 1 \end{pmatrix}, \qquad R_2 = -\frac{1}{2} \begin{pmatrix} -1 & \omega\tau & \frac{\omega^2}{\tau} \\ \frac{\tau}{\omega} & \tau^{-1} & \omega \\ \frac{\omega}{\tau} & \omega^2 & -\tau \end{pmatrix},$$

$$R_3 = \begin{pmatrix} -1 & 0 & 0 \\ 0 & 1 & 0 \\ 0 & 0 & 1 \end{pmatrix},$$

where $\omega = \exp(2\pi i/3)$ and $\tau = (1 + \sqrt{5})/2$. The Valentiner group has order 2,160 and the corresponding projective group in $\mathrm{PGL}(3, \mathbb{C})$ is isomorphic to A_6. The Valentiner arrangement is given by the $k = 45$ lines fixed by the involutions of A_6. We have $t_3 = 120$, $t_4 = 45$, $t_5 = 36$, and all other $t_r = 0$. For each $i = 1, \ldots, k$, we have $t_3^{(i)} = 8$, $t_4^{(i)} = 4$, $t_5^{(i)} = 4$, so that $3\tau = k + 3$ as required. As we saw in Chapter 2, §2.7, the groups A_5 and G_{168} can be represented on a vector space of dimension 3. This is not true, however, of the group A_6, which operates on the projective plane but has only a three dimensional projective representation.

We have therefore checked the first part of the following result, proved by Höfer in his thesis (see [11]):

Theorem 5.1 *For all the above arrangements, we have $3\tau = \tau_i = k + 3$ for all lines $L_i, i = 1, \ldots, k$. Moreover, for $k > 3$, the matrix R is positive semi-definite.*

The excluded example with $k = 3$ is the triangle, given by $x_0 x_1 x_2 = 0$, for which the matrix R is negative semi-definite.

Except in the cases of either the $\overline{\mathrm{Ceva}}(q)$, $q \geq 2$, and the extended Hesse arrangement, where Höfer checks that R is positive semi-definite by other methods, we can give Höfer's proof that R is positive semi-definite as follows. Notice that the space in $\mathbb{C}^k$ defined by those $(x_i)_{i=1}^k \in \mathbb{C}^k$ with $\sum_i x_i = 0$ is orthogonal to the vector with all $x_i = 1$. For the configurations considered in this section, we have already checked that the vector with $x_i = 1$ for all i is in the kernel of R. Let R' be the matrix defined by

$$R = 3R' - \begin{pmatrix} 1 & 1 & \cdots & 1 \\ 1 & 1 & \cdots & 1 \\ \cdot & \cdot & \cdots & \cdot \\ \cdot & \cdot & \cdots & \cdot \\ \cdot & \cdot & \cdots & \cdot \\ 1 & 1 & \cdots & 1 \end{pmatrix}, \tag{5.25}$$

where all entries of the matrix on the right-hand side are 1. Then, the matrix R' has entries

$$R'_{ii} = \sigma_i - 1,$$

$$R'_{il} = 1, \; L_i \cap L_l = \{\text{regular point}\}, \; i \neq l,$$

$$R'_{il} = 0, \; L_i \cap L_l = \{\text{singular point}\}, \; i \neq l. \tag{5.26}$$

The matrix $3R'$ acts in the same way as the matrix R on a vector with $\sum_i x_i = 0$. We now apply the following result of Gershgorin.

Theorem 5.2 *Let $A = (a_{ij})$ be a complex $k \times k$ matrix. Then, all the eigenvalues of A are contained in the union of k closed disks with centers a_{ii} and radii $\sum_{i \neq j} |a_{ij}|$, respectively.*

Proof. Recall the eigenvalue equation:

$$Ax = \lambda x, \qquad x \in \mathbb{C}^k \setminus \{0\}.$$

Let i satisfy $|x_i| = \max_j |x_j|$. Then, by the eigenvalue equation,

$$a_{ii} x_i + \sum_{j \neq i} a_{ij} x_j = \lambda x_i.$$

Dividing by x_i, which is nonzero, we have

$$\lambda - a_{ii} = \sum_{j \neq i} a_{ij} \frac{x_j}{x_i}.$$

Therefore,

$$|\lambda - a_{ii}| \leq \sum_{j \neq i} |a_{ij}| |\frac{x_j}{x_i}|.$$

But $|\frac{x_j}{x_i}| \leq 1$, which gives the result.

Applying the above result, we see that the matrix R' is positive definite if $t_2^{(i)} < \sigma_i - 1$ for all $i = 1, \ldots, k$. Now, $\tau_i = \sigma_i + t_2^{(i)}$, so if $3\tau_i = k + 3$, we have

$$(\sigma_i - 1) = \frac{k}{3} - t_2^{(i)}.$$

In this case, for R' to be positive definite, it suffices that

$$t_2^{(i)} < \frac{k}{6},$$

and for R' to be positive semi-definite, equality suffices. This argument shows that R' is positive definite for the arrangements Ceva(q), $q \geq 3$, A_5 (icosahedral), G_{168}, and A_6. It shows that R' is positive semi-definite for Ceva(2), and, therefore, also for $\overline{\text{Ceva}}(1)$, as well as for the Hesse arrangement. However, it does not enable us to determine the rank of the kernel of R' for these arrangements. Of course, if $t_2 = 0$ the matrix R' is diagonal, namely, $\frac{k}{3} \cdot I_k$, and clearly positive definite.

If the matrix R' is positive definite, then the line arrangement can only annihilate R when all the lines have the same weight. If the matrix R' has a nontrivial kernel, there may be other possibilities. We shall return to this in §5.5.

Following [11], 5.6, and 5.7, we complete this discussion to obtain the positive semi-definiteness of R' and the rank of its kernel in all cases.

For the Hesse arrangement, R' is a 12×12 matrix consisting of four blocks of type

$$\begin{pmatrix} 2 & 1 & 1 \\ 1 & 2 & 1 \\ 1 & 1 & 2 \end{pmatrix}$$

corresponding to the four triangles of the arrangement. Hence, R' is positive definite.

For $\overline{\text{Ceva}}(q)$, the matrix R' is as follows.

$$\begin{pmatrix} 1 & 0 & 0 & 1_q & 0_q & 0_q \\ 0 & 1 & 0 & 0_q & 1_q & 0_q \\ 0 & 0 & 1 & 0_q & 0_q & 1_q \\ 1_q^t & 0_q^t & 0_q^t & q1_{q,q} & 0_{q,q} & 0_{q,q} \\ 0_q^t & 1_q^t & 0_q^t & 0_{q,q} & q1_{q,q} & 0_{q,q} \\ 0_q^t & 0_q^t & 1_q^t & 0_{q,q} & 0_{q,q} & q1_{q,q} \end{pmatrix}. \qquad (5.27)$$

Here, 1_q is the row vector of length q all of whose entries equal 1, and 0_q is the row vector of length q all of whose entries are 0. Further, $1_{q,q}$ is the $q \times q$ identity matrix and $0_{q,q}$ is the $q \times q$ zero matrix. The matrix displayed above is clearly positive semi-definite, and the kernel of R' has rank 3. It consists of those elements of the kernel of R assigning equal weights to all the lines passing through one of the $(q + 2)$-fold points. As there are three such $(q + 2)$-fold points, the kernel has rank 3. For example, the following vector is in the kernel of R', as well as R:

$$(q, 0, 0, -1_q, 0_q, 0_q).$$

For the extended Hesse arrangement, the matrix R' is of the form

$$R' = \begin{pmatrix} 4 1_{12,12} & {}^t A \\ A & 3 1_{9,9} \end{pmatrix},$$

where the 9×12 matrix A describes the intersection behavior between a line of $C_{12} = 0$ and a line of $C_9 = 0$. Each line of $C_{12} = 0$ passes through three regular points, and three four-fold points and two five-fold points of the extended Hesse arrangement. It meets the nine lines of $C_9 = 0$ in the regular points and the five-fold points. Namely, a line of $C_{12} = 0$ meets one line of $C_9 = 0$ in a regular point and three lines of $C_9 = 0$ in a five-fold point. Each line of $C_9 = 0$ has four regular and four five-fold points, where it meets the twelve lines of $C_{12} = 0$. Therefore, each column of A contains the entry 1 three times, and each row contains the entry 1 four times. It follows that the entry 1 appears altogether thirty-six times in A: these entries correspond to the thirty-six double points of the extended Hesse arrangement. It follows that the row vector of length 21 given by

$$\left(-\tfrac{3}{4} 1_{12}, \, 1_9 \right)$$

is in the kernel of R' as well as the kernel of R. We can check that

$$R' = \begin{pmatrix} 2 1_{12} & 0 \\ A/2 & {}^t C/2 \end{pmatrix} \begin{pmatrix} 2 1_{12} & {}^t A/2 \\ 0 & C/2 \end{pmatrix}$$

if and only if

$$^t C \cdot C = B,$$

where the coefficients B_{ij} of B satisfy $B_{ij} = 8$ for $i = j$, and $B_{ij} = -1$ for $i \neq j$. The matrix B has a kernel of rank 1 generated by $(1, 1, \ldots, 1)$ and is positive semi-definite. Both facts follow from Theorem 5.2. Therefore, a lower triangular matrix C as above exists, the matrix R' is positive semi-definite, and C and R' each have a kernel of rank 1. The kernel of R consists of the vectors having equal coordinates on the twelve Hesse lines and equal coordinates on the nine Ceva lines. We therefore arrive at the following:

Theorem 5.3 *The known arrangements with $k > 3$ and $3\tau = k + 3$ have a matrix R' that is positive definite except for* Ceva(2), *or, equivalently, for*

$\overline{\text{Ceva}}(1)$, for $\overline{\text{Ceva}}(q)$, and for the extended Hesse arrangement. In these cases, the matrix R' is positive semi-definite and has kernel of rank 3 for $\overline{\text{Ceva}}(q)$, $q \geq 1$, and of rank 1 for the extended Hesse arrangement.

5.4 BLOW-UP OF A SINGULAR INTERSECTION POINT

Consider an r-fold intersection point ($r \geq 3$) of a line arrangement on $\mathbb{P}_2$ and the blow-up X of $\mathbb{P}_2$ at this point. We therefore have r lines on X that are proper transforms of the original lines of the arrangement on $\mathbb{P}_2$ that pass through the blown-up point. We now assign ramification indices $n_i \geq 2$, $i = 1, \ldots, r$, to these proper transforms. To the blown-up point, that is, the resulting exceptional divisor E with $E \cdot E = -1$, we assign the ramification $m \geq 1$. Consider a good covering Y of degree N over X, satisfying the conditions given by Definition 5.1, §5.1. Over E, we have a smooth curve $\widetilde{E}$ on Y. As $m\widetilde{E}$ is the lift of the divisor E, we have

$$(m\widetilde{E}) \cdot (m\widetilde{E}) = -N,$$

which we write as

$$\widetilde{E} \cdot \widetilde{E} = -\frac{N}{m^2}. \tag{5.28}$$

Moreover,

$$e(\widetilde{E}) = -\frac{N}{m} \left((r-2) - \sum_{i=1}^{r} \frac{1}{n_i} \right). \tag{5.29}$$

By Definition 5.2, we have

$$\text{prop}(\widetilde{E}) = 2\widetilde{E} \cdot \widetilde{E} - e(\widetilde{E}).$$

Equations (5.28) and (5.29) give

$$\text{prop}(\widetilde{E}) = 2\widetilde{E} \cdot \widetilde{E} - e(\widetilde{E})$$

$$= -\frac{2N}{m^2} + \frac{N}{m} \left((r-2) - \sum_{i=1}^{r} \frac{1}{n_i} \right)$$

$$= \frac{N}{m} \left(\frac{-2}{m} + (r-2) - \sum_{i=1}^{r} \frac{1}{n_i} \right). \tag{5.30}$$

We can rewrite (5.30) as

$$\text{prop}(\widetilde{E}_j) = \frac{N}{m} P_j = \frac{N}{m} \left(2z_j - 4 + \sum_{i=1}^{r} x_i D_i \cdot E_j \right).$$

The right-hand side of (5.30) can be recovered by substituting into this last formula

$$E_j = E, \qquad z_j = 1 - \frac{1}{m}, \qquad x_i = 1 - \frac{1}{n_i}, \qquad i = 1, \dots, r.$$

The blown-up point will contribute nothing to $\text{Prop}(Y)$ when $\text{prop}(\widetilde{E}) = 0$ (see (5.19)). By (5.30), we have $\text{prop}(\widetilde{E}) = 0$ for those $n_i \geq 2$, $i = 1, \dots, r$, and $m \geq 1$ satisfying

$$\frac{2}{m} + \sum_{i=1}^{r} \frac{1}{n_i} = r - 2. \qquad (5.31)$$

Equation (5.31) has only finitely many solutions. Those were found explicitly by Höfer [59], and for each of them $r \leq 8$. For $r = 3$, there are eighty-seven solutions, for $r = 4$ there are twenty-seven, for $r = 5$ there are 150, for $r = 6$ there are eighteen, for $r = 7$ there are three, and for $r = 8$ there is one. For example, for $r = 3$ we have the solution $n_i = 5$, $i = 1, 2, 3$ and $m = 5$. We do not list all the solutions to (5.31) in this book, as we are only interested in weights that can be assigned to the list of line arrangements introduced in §5.3 and that give rise to finite covers by ball quotients. This places additional restrictions on the possible solutions to (5.31). For each arrangement, the possible weights, computed in §5.5, are listed in the tables given in this chapter. In Appendix B, we are exclusively interested in weighted line arrangements that give rise to Kummer coverings.

The list of solutions to (5.31) can be extended if we admit the conditions $m < 0$ and $m = \infty$. The condition $m < 0$ implies that

$$\sum_{i=1}^{r} \frac{1}{n_i} > r - 2. \qquad (5.32)$$

Only the case $r = 3$ can occur, as we have excluded $r = 2$. In this way we recover the list of spherical, or finite, triangle groups of Theorem 2.1, Chapter 2; namely, we have

$$(n_1, n_2, n_3) = (2, 3, 3), \ (2, 3, 4) \ (2, 3, 5), \ \text{and} \ (2, 2, p), \ 2 \leq p < \infty.$$

If we admit $m = \infty$, we have the Euclidean r-gon condition

$$\sum_{i=1}^{r} \frac{1}{n_i} = r - 2. \tag{5.33}$$

For $r = 3$, we thereby recover the corresponding Euclidean triangle groups of signature $(2, 4, 4)$, $(3, 3, 3)$, and $(2, 3, 6)$; and for $r = 4$, we have $(2, 2, 2, 2)$. Together these give the entire list of solutions.

We now assume that, for each exceptional divisor E_j on X corresponding to a blown-up point p_j of the arrangement on $\mathbb{P}_2$, there exists an integer m_j with $1 \leq |m_j| < \infty$ or $m_j = \infty$ such that

$$\frac{2}{m_j} + \sum_{i=1}^{r} \frac{1}{n_i} = r - 2. \tag{5.34}$$

Moreover, we take the ramification index over E_j to be $|m_j|$ if $m_j \neq \infty$, and to be an arbitrary natural number m'_j if $m_j = \infty$. Recall Höfer's formula (5.11),

$$\mathrm{Prop}(Y)/N = (3c_2(Y) - c_1^2(Y))/N = \frac{1}{4} \sum P_j^2 + \frac{1}{4} \sum R_{il} x_i x_l.$$

By our assumptions, the sum $\frac{1}{4} \sum P_j^2$ need only be extended over the E_j with $m_j < 0$ or $m_j = \infty$. Let us begin with the spherical case $m_j < 0$. Consider a given blown-up point and write $m_j = m$ and $E_j = E$. The inverse image, $\widetilde{E}$, of E on Y satisfies (5.28). Moreover, by (5.29), with m replaced by $-m$, and by (5.34),

$$e(\widetilde{E}) = \frac{-N}{m} \left(\frac{1}{n_1} + \ldots + \frac{1}{n_r} - (r - 2) \right)$$

$$= \frac{-N}{m} \left(\frac{-2}{m} \right) = \frac{2N}{m^2}. \tag{5.35}$$

It follows by (5.28) and (5.35) that

$$\mathrm{prop}(\widetilde{E}) = \frac{N}{m^2}(-4),$$

giving

$$\frac{1}{4} P_j^2 = \frac{1}{4} \left(\frac{m}{N} \mathrm{prop}(\widetilde{E}) \right)^2 = \frac{4}{m^2}. \tag{5.36}$$

From the value of $e(\widetilde{E})$ in (5.35), we conclude that $\widetilde{E}$ consists of N/m^2 smooth disjoint rational curves. The value of the self-intersection number of $\widetilde{E}$

in (5.28) makes it clear that all of these curves are exceptional. We now blow down all the exceptional curves on Y obtained in this way and denote the resulting surface by Y'. Each blow-down decreases Prop by 4. Hence,

$$\text{Prop}(Y')/N = \frac{1}{4} \sum_{m_j = \infty} P_j^2 + \frac{1}{4} \sum R_{il} x_i x_l, \qquad (5.37)$$

where the first sum on the right-hand side is only taken over those E_j with $m_j = \infty$. Consider one such E_j with $m_j = \infty$, and set $E_j = E$ and $m'_j = m'$. Then, by (5.28), with m replaced by m',

$$\widetilde{E} \cdot \widetilde{E} = \frac{-N}{m'^2}, \quad e(\widetilde{E}) = 0. \qquad (5.38)$$

Therefore,

$$\text{prop}(\widetilde{E}) = -\frac{2N}{m'^2},$$

so

$$\frac{1}{4} P_j^2 = \frac{1}{4} \left(\frac{m'}{N} \frac{2N}{m'^2} \right)^2 = \frac{1}{m'^2}. \qquad (5.39)$$

The inverse image, $\widetilde{E}$, of E on Y consists of disjoint irreducible elliptic curves. The sum of their (negative) self-intersection numbers is $-N/m'^2$. Therefore, it follows by (5.37) and (5.39) that

$$\text{Prop}(Y') = \frac{N}{4} \sum_{i,l} R_{il} x_i x_l - \sum C_r \cdot C_r. \qquad (5.40)$$

The second sum is extended over all elliptic curves C_r obtained in the above way, when we take into account all P_j with $m_j = \infty$.

If we have a smooth elliptic curve C on a complex compact surface Y with $C \cdot C < 0$, we define $c_1(Y, C)$ by

$$c_1(Y, C) = c_1(Y) - \alpha C,$$

for an $\alpha > 0$ to be chosen below. The right hand side makes sense using the Poincaré isomorphism. We require

$$c_1(Y, C) \cdot C = 0$$

in order that $c_1(Y, C)$ correspond to a divisor on the surface $Y \setminus C$. According to the adjunction formula,

$$c_1(Y) \cdot C - C \cdot C = e(C) = 0.$$

We therefore choose $\alpha = 1$, so that

$$c_1(Y, C) = c_1(Y) - C. \tag{5.41}$$

Therefore,

$$c_1^2(Y, C) = c_1^2(Y) - C \cdot C.$$

We define

$$c_2(Y, C) = c_2(Y \setminus C) = c_2(Y). \tag{5.42}$$

The Chern classes $c_1(Y, C)$ and $c_2(Y, C)$ are special cases of the logarithmic Chern classes discussed at the beginning of Chapter 6. Similar formulas hold when we have finitely many disjoint elliptic curves C_r with negative self-intersection. Writing $C = \sum_r C_r$ and letting

$$\mathrm{Prop}(Y, C) = 3c_2(Y, C) - c_1^2(Y, C), \tag{5.43}$$

we have

$$\mathrm{Prop}(Y) = \mathrm{Prop}(Y, C) - C \cdot C. \tag{5.44}$$

Hence, from the formula (5.40) for $\mathrm{Prop}(Y')$, we deduce that

$$\mathrm{Prop}(Y', C) = \frac{N}{4} \sum_{i,l} R_{il} x_i x_l. \tag{5.45}$$

5.5 POSSIBILITIES FOR THE ASSIGNED WEIGHTS

If the matrix R of an arrangement is positive semi-definite, then, by Höfer's formula (5.11), the number $\mathrm{Prop}(Y)$ cannot be 0 if one of the $\widetilde{E}_j$ is a union of rational or elliptic curves, or if we have $P_j \neq 0$ for one of the other $\widetilde{E}_j$. Moreover, the only way to ensure that $\mathrm{Prop}(Y', C) = 0$, where $\mathrm{Prop}(Y', C)$, is as in §5.4, is to annihilate all the P_j by the condition

$$\frac{2}{m_j} + \sum_{P_j \in L_i} \frac{1}{n_i} = r - 2, \tag{5.46}$$

and, also, to require that the vector $(x_1, x_2, \ldots, x_k)$, where $x_i = 1 - \frac{1}{n_i}$, be in the kernel of R. For the time being, let us make the following assumption:

(†) We allow only $n_i \geq 2$ and finite, but permit $m_j \geq 1$, $m_j < 0$ or $m_j = \infty$.

Condition (†) means we are looking for finite covers of X, the projective plane blown up at the singular points of the line arrangement, which have finite positive nontrivial ramifications over the proper transforms of the lines. This is a natural geometric condition if one is focusing on finite covers of blow-ups of the arrangement that are ball quotients.

When $m_j < 0$, equation (5.46) is equivalent to the fact that the curves $\widetilde{E}_j$ are rational and can be blown down. We have $m_j = \infty$ in (5.46) if and only if $\widetilde{E}_j$ is elliptic. The elliptic curves are to be omitted.

We now consider the arrangements of §5.3, except the triangle and Ceva(1). For each of these arrangements, the matrix R is positive semi-definite. We shall give all weights satisfying (†) and leading to Prop$(Y', C) = 0$.

For the Hesse, icosahedral, G_{168}, A_6, and Ceva(q), ($q \geq 3$) arrangements, the corank of R is 1. For the extended Hesse, the corank is 2. For the $\overline{\text{Ceva}}(q)$ ($q \geq 1$), the corank is 4. Recall that $\overline{\text{Ceva}}(1)$ and Ceva(2) are the complete quadrilateral arrangement. As we shall see, for both Ceva(q) and $\overline{\text{Ceva}}(q)$, the choice of allowable weights depends on q.

We first consider those arrangements for which the matrix R has corank 1. As these arrangements satisfy $3\tau = k + 3$, we know that $(1, 1, \ldots, 1)$ ($x_i = 1$, $i = 1 \ldots, k$) generates the kernel of R. We are thus in the case of equal weights, $n_i = n$, for $i = 1, \ldots, k$. By (5.46), this implies that the diophantine condition at a blown-up $r \geq$ threefold point with $m_j = m$ becomes

$$\frac{r}{n} + \frac{2}{m} = r - 2. \tag{5.47}$$

We now refer the reader to the tables given in this chapter. Table 5.1 displays the complete list of solutions to (5.47), subject to our restrictions on n, m, and r (namely (†) above and $r \geq 3$).

Hesse arrangement: There are only twofold and fourfold intersection points. For $r = 4$, the cases $n = 2, 3, 4$ occur in Table 5.1, from which we deduce the three possibilities for this arrangement given in Table 5.2.

Icosahedral arrangement: In this case, there are only twofold, threefold and fivefold intersection points. For $r = 3$, the cases $n = 2, 3, 4, 5, 6, 9$ occur in Table 5.1, and for $r = 5$, the cases $n = 2, 5$ occur in Table 5.1. As we have threefold and fivefold points, only the cases $n = 2$ and $n = 5$ are possible for the icosahedral arrangement. The two possibilities are given in Table 5.3.

G_{168} arrangement: There are only threefold and fourfold intersection points, so we have the three possibilities given in Table 5.4.

TABLE 5.1 Solutions to (5.47)

n	r	m
2	5	4
2	6	2
2	8	1
2	3	-4
2	4	∞
3	4	3
3	6	1
3	3	∞
4	3	8
4	4	2
5	3	5
5	5	1
6	3	4
9	3	3

TABLE 5.2 Hesse arrangement

n	r	m
2	4	∞
3	4	3
4	4	2

TABLE 5.3 Icosahedral arrangement

n	r	m
2	3	-4
2	5	4
5	3	5
5	5	1

TABLE 5.4 G_{168} arrangement

n	r	m
2	3	-4
2	4	∞
3	3	∞
3	4	3
4	3	8
4	4	2

TABLE 5.5 A_6 arrangement

n	r	m
2	3	-4
2	4	∞
2	5	4

TABLE 5.6 Ceva(3)

n	r	m
2	3	-4
3	3	∞
4	3	8
5	3	5
6	3	4
9	3	3

TABLE 5.7 Ceva(4)

n	r	m
2	3	-4
2	4	∞
3	3	∞
3	4	3
4	3	8
4	4	2

TABLE 5.8 Ceva(5)

n	r	m
2	3	-4
2	5	4
5	3	5
5	5	1

A_6 **arrangement:** There are only threefold, fourfold, and fivefold intersection points, so we have must have $n = 2$, and the only possibilities are given in Table 5.5.

Ceva(q) ($q \geq 3$) arrangement: There are only threefold and q-fold points. Clearly the possibilities depend on q. For $q = 3, 4, 5, 6, 8$, the possibilities are given in Tables 5.6 to 5.10. There are no possibilities for $q = 7$ and for $q \geq 9$.

TABLE 5.9 Ceva(6)

n	r	m
2	3	-4
2	6	2
3	3	∞
3	6	1

TABLE 5.10 Ceva(8)

n	r	m
2	3	-4
2	8	1

Remark. For $q = 3$, later we will also consider the cases where $n = \infty, r = 3$, and $m = 2$, and $n = -3$, $r = 3$, and $m = 1$. The latter case corresponds to a K_3 surface, so, in particular, the canonical divisor K vanishes. For $q = 4$, later we will also consider the cases where $n = \infty$, $r = 4$ and $m = 1$, which gives a vanishing coefficient of K, and $n = \infty$, $r = 3$, and $m = 2$. For $q = 5$, the case $n = 5$, $r = 5$, $m = 1$ gives a vanishing coefficient of K. For $q = 6$, the case $n = 3$, $r = 6$, $m = 1$ gives a vanishing coefficient of K. For $q = 8$, the case $n = 2, r = 8, m = 1$ gives a vanishing coefficient of K.

We now consider the arrangements for which the corank of R is greater than 1. We still assume (†) and $r \geq 3$.

Extended Hesse arrangement (corank(R) = 2): We only have twofold, fourfold, and fivefold points. The arrangement consists of the lines of the Hesse arrangement together with those of the Ceva(3) arrangement. According to §5.3, the lines of the Hesse arrangement must have equal weights n_H and those of the Ceva(3) arrangement must have equal weights n_C. At the fourfold points, from (5.47) we obtain the diophantine restriction

$$\frac{4}{n_H} + \frac{2}{m} = 2,$$

which has the following table of solutions:

n_H	r	m
2	4	∞
3	4	3
4	4	2

TABLE 5.11 Extended Hesse arrangement

n_H	n_C	m_4	m_5
2	2	∞	4
2	3	∞	2
3	9	3	1
4	2	2	2
4	6	2	1

At the fivefold points, from (5.46) we obtain the diophantine restriction

$$\frac{2}{n_H} + \frac{3}{n_C} + \frac{2}{m} = 3,$$

which has the following table of solutions:

n_H	n_C	r	m
2	2	5	4
2	3	5	2
3	9	5	1
4	2	5	2
4	6	5	1

The possible resulting weights n_H and n_C are given in Table 5.11. We give there the value m_4 of m at the fourfold point, and its value m_5 at the fivefold point.

$\overline{\text{Ceva}}(q)$ arrangement, $q \geq 1$, (corank(R) = 4):

If x_1, x_2, x_3 are homogeneous coordinates for $\mathbb{P}_2$, recall that the $\overline{\text{Ceva}}(q)$ arrangement has the equation

$$x_1 x_2 x_3 (x_1^q - x_2^q)(x_3^q - x_2^q)(x_3^q - x_1^q) = 0,$$

and that for $q \neq 1$ it has $k = 3q + 3$ lines, with $t_2 = 3q$, $t_3 = q^2$, $t_{q+2} = 3$, and $t_r = 0$ otherwise. When $q = 1$, we have the complete quadrilateral arrangement $\overline{\text{Ceva}}(1)$, with $k = 6$ lines and $t_2 = 3$, $t_3 = 4$, with $t_r = 0$ otherwise. In all cases, the corank of R is 4.

We begin by studying $\overline{\text{Ceva}}(1)$. Recall from Chapter 3, §3.3, and §5.2 of this chapter, that we can parametrize the kernel of R by parameters $\mu_1, \mu_2, \mu_3, \mu_4$ and that we set

$$x_{12} = \mu_1 + \mu_2, \quad x_{13} = \mu_1 + \mu_3, \quad x_{23} = \mu_2 + \mu_3, \quad x_{14} = \mu_1 + \mu_4,$$

$$x_{24} = \mu_2 + \mu_4, \quad x_{34} = \mu_3 + \mu_4,$$

where $x_{\alpha\beta} = 1 - \frac{1}{n_{\alpha\beta}}$ and the $n_{\alpha\beta}$ are the weights associated to the six lines, $\alpha, \beta = 1, 2, 3, 4$. Associated to the blown-up points, we have the ramifications $n_{0\beta}$, with $x_{0\beta} = 1 - \frac{1}{n_{0\beta}} = \mu_0 + \mu_\beta, \beta = 1, 2, 3, 4$, where $\sum_{i=0}^{4} \mu_i = 2$.

In his 1896 dissertation, R. Le Vavasseur, a student of Picard, found all solutions to the equations

$$1 - \frac{1}{n_{\alpha\beta}} = \mu_\alpha + \mu_\beta, \quad n_{\alpha\beta} \in \mathbb{Z} \setminus \{0\} \quad \text{or} \quad n_{\alpha\beta} = \infty \quad \text{and} \quad \mu_\alpha \in \mathbb{Q},$$

$\alpha, \beta = 0, \ldots, 4$. The complete list of solutions, up to permutation for the quintuplets $(\mu_0, \ldots, \mu_4)$, is given in Table 5.13, which is taken from [11], p. 199. In Table 5.13, we use the notation $\mu_i = \alpha_i/d, d \geq 1$, and under DM we give the numbering in the list [33]. The twenty-eight cases, 1 through 27 and 31, can be characterized by the fact that the weight 1 does not occur, that is, $\mu_\alpha + \mu_\beta \neq 0$ for $\alpha \neq \beta$. The cases 1 through 27 also satisfy $0 < \mu_i < 1$, $i = 0, \ldots, 4$. In Table 5.12, we rewrite 1 through 27 and 31 in terms of the weights of the $D_{\alpha\beta}$. Here, the $n_i, i = 1, \ldots, 6$, are replaced by the $n_{\alpha\beta}$ (where $\alpha < \beta$ and $\alpha, \beta = 1, \ldots 4$.) The m_j are replaced by n_{01}, n_{02}, n_{03}, and n_{04}. This list is taken from [11], p. 201. Strictly speaking, for us the cases 20 and 23 in Table 5.12 have to be omitted because (†) is not satisfied, whereas for the others it is satisfied. At the same time, this is not very consequential, because in the examples we can interchange the role of all the lines $D_{\alpha\beta}$. Moreover, we shall see later that we can generalize the theory in such a way that lines of the arrangement may have the weight ∞ or negative weights. We cannot treat 20 or 23 by blowing down the disjoint lines D_{01}, D_{02}, D_{12}. This leads to $\mathbb{P}_1 \times \mathbb{P}_1$, and not to $\mathbb{P}_2$, because D_{34} has self-intersection number 2 after these blow-downs.

Le Vavasseur gave a geometric interpretation of a large part of his list. The numbers 1 through 27 of Table 5.12 lead to ball quotients, whereas 31 satisfies (†), but leads to an abelian surface.

We shall now study $\overline{\text{Ceva}}(q)$, for $q > 1$. Under the map

$$\rho_q : [x_0 : x_1 : x_2] \mapsto [x_0^q : x_1^q : x_2^q]$$

of $\mathbb{P}_2$ to itself, the complete quadrilateral $\overline{\text{Ceva}}(1)$ lifts to $\overline{\text{Ceva}}(q)$. That is, if $\rho_q(x)$ lies on $\overline{\text{Ceva}}(1)$, then x lies on $\overline{\text{Ceva}}(q)$, for $x \in \mathbb{P}_2$. However, a covering of $\mathbb{P}_2$ branched along $\overline{\text{Ceva}}(1)$ with weights $n_{\alpha\beta}$ satisfying (5.46) may not lift to a covering of $\overline{\text{Ceva}}(q)$ with weights $n_{\alpha\beta}^{(q)}$ satisfying (5.46). This places conditions on the $n_{\alpha\beta}$. Let

$$n_{14}, \ n_{24}, \ n_{34}$$

be the respective weights associated to the outside lines

$$x_1 = 0, \ x_2 = 0, \ x_3 = 0$$

TABLE 5.12 $\overline{\text{Ceva}}(1)$. (31 gives an abelian surface)

Nr		n_{01}	n_{02}	n_{03}	n_{04}	n_{12}	n_{13}	n_{14}	n_{23}	n_{24}	n_{34}	
1		5	5	5	5	5	5	5	5	5	5	†
2		8	8	8	8	4	4	4	4	4	4	†
3		−4	8	8	8	8	8	8	2	2	2	†
4		−8	−8	−8	8	4	4	2	4	2	2	†
5		9	9	9	3	9	9	3	9	3	3	†
6		−10	−10	−10	5	5	5	2	5	2	2	†
7		6	6	6	4	6	6	4	6	4	4	†
8		12	12	6	6	6	4	4	4	4	3	†
9		12	12	12	4	6	6	3	6	3	3	†
10		−6	12	12	4	12	12	4	3	2	2	†
11		−12	−12	12	12	6	3	3	3	3	2	†
12		−12	−12	−12	4	6	6	2	6	2	2	†
13		−4	12	12	12	6	6	6	2	2	2	†
14		−4	−12	−12	−12	3	3	3	2	2	2	†
15		15	15	15	5	5	5	3	5	3	3	†
16		−18	−18	−18	3	9	9	2	9	2	2	†
17		−4	20	20	20	5	5	5	2	2	2	†
18		24	24	24	8	4	4	3	4	3	3	†
19		∞	∞	∞	∞	3	3	3	3	3	3	†
20		∞	∞	4	4	∞	4	4	4	4	2	
21		−4	∞	∞	∞	4	4	4	2	2	2	†
22		∞	6	6	6	6	6	6	3	3	3	†
23		∞	∞	6	3	∞	6	3	6	3	2	
24		−6	∞	∞	6	6	6	3	3	2	2	†
25		−6	−6	−6	∞	3	3	2	3	2	2	†
26		∞	12	12	12	4	4	4	3	3	3	†
27		−12	∞	6	6	12	4	4	3	3	2	†
31	X	−4	−4	−4	−4	2	2	2	2	2	2	†

of $\overline{\text{Ceva}}(1)$, and let

$$n_{12}, \ n_{23}, \ n_{13}$$

be the respective weights associated to the lines

$$x_1 = x_2, \ x_2 = x_3, \ x_1 = x_3.$$

A covering of $\mathbb{P}_2$ branched along $\overline{\text{Ceva}}(1)$ with weights $n_{\alpha\beta}$ lifts to a covering of $\overline{\text{Ceva}}(q)$ only if n_{14}, n_{24}, n_{34} are divisible by q. Let $n_{14}^{(q)}, n_{24}^{(q)}, n_{34}^{(q)}$ be the weights

TABLE 5.13 Le Vavasseur's list for $\overline{\text{Ceva}}(1)$

Nr		d		α_0	α_1	α_2	α_3	α_4		DM
1		5		2	2	2	2	2		4
2		8		4	3	3	3	3		9
3		8		5	5	2	2	2		10
4		8		6	3	3	3	1		11
5		9		4	4	4	4	2		12
6		10		7	4	4	4	1		13
7		12		5	5	5	5	4		14
8		12		6	5	5	4	4		15
9		12		6	5	5	5	3		16
10		12		7	7	4	4	2		19
11		12		8	5	5	3	3		20
12		12		8	5	5	5	1		21
13		12		8	7	3	3	3		22
14		12		10	5	3	3	3		23
15		15		8	6	6	6	4		24
16		18		11	8	8	8	1		25
17		20		14	11	5	5	5		26
18		24		14	9	9	9	7		27
19		3		2	1	1	1	1		1
20		6		3	3	3	2	1		6
21		4		3	2	1	1	1		3
22		6		3	3	2	2	2		5
23		4		2	2	2	1	1		2
24		6		4	3	2	2	1		7
25		6		5	2	2	2	1		8
26		12		7	5	4	4	4		17
27		12		7	6	5	3	3		18
28		d		$d+1$	$d-1$	0	0	0		
29		3		2	2	2	0	0		
30		4		3	3	2	0	0		
31		4		4	1	1	1	1		
32		6		5	4	3	0	0		
33		1		1	1	0	0	0		
34		2		2	1	1	0	0		
35		2		1	1	1	1	0		
36		2		2	1	1	1	-1		
37		3		2	2	2	2	-2		
38		6		4	4	4	1	-1		

associated to the outside lines $x_1 = 0$, $x_2 = 0$, $x_3 = 0$ of $\overline{\text{Ceva}}(q)$, and let $n_{12}^{(q)}$, $n_{23}^{(q)}$, $n_{13}^{(q)}$ be the weights associated to the respective groups of q lines $x_1^q = x_2^q$, $x_2^q = x_3^q$, $x_1^q = x_3^q$. Then,

$$n_{14}^{(q)} = \frac{n_{14}}{q}, \quad n_{24}^{(q)} = \frac{n_{24}}{q}, \quad n_{34}^{(q)} = \frac{n_{34}}{q},$$

$$n_{12}^{(q)} = n_{12}, \quad n_{23}^{(q)} = n_{23}, \quad n_{13}^{(q)} = n_{13}.$$

We rewrite the above equations in terms of the parameters μ_1, μ_2, μ_3, μ_4 of the four dimensional kernel of the matrix R for $\overline{\text{Ceva}}(1)$. We have

$$x_{12} = \mu_1 + \mu_2, \quad x_{13} = \mu_1 + \mu_3, \quad x_{23} = \mu_2 + \mu_3, \quad x_{14} = 1 - q + q(\mu_1 + \mu_4),$$

$$x_{24} = 1 - q + q(\mu_2 + \mu_4), \quad x_{34} = 1 - q + q(\mu_3 + \mu_4), \quad (5.48)$$

where $x_{\alpha\beta} = 1 - \frac{1}{n_{\alpha\beta}^{(q)}}$. This is equivalent to the formulas for the $n_{\alpha\beta}^{(q)}$ in terms of the $n_{\alpha\beta}$ since $n_{\alpha\beta} = (1 - \mu_\alpha - \mu_\beta)^{-1}$. In this way, we express the $x_{\alpha\beta}$ in terms of the four independent parameters μ_i, $i = 1, 2, 3, 4$. We check that this corresponds to a parametrization of the kernel of R for $\overline{\text{Ceva}}(q)$ using (5.25) and (5.27). From (5.27), the value of the dot product of each line of $3R'$ with the vector of length $3q + 3$ corresponding to $x_{\alpha\beta}$ (x_{14}, x_{24}, x_{34} appearing once, and x_{12}, x_{23}, x_{13} each appearing q times) is

$$3(1 - q + q(\mu_1 + \mu_2 + \mu_3 + \mu_4)),$$

which equals the sum

$$q(x_{12} + x_{13} + x_{23}) + x_{14} + x_{24} + x_{34}$$

of all components of the vector. Therefore, from (5.25), we see that the vector of length $3q + 3$ corresponding to $x_{\alpha\beta}$ is indeed in the kernel of R.

Recall that $\overline{\text{Ceva}}(q)$ has three $(q + 2)$-fold points. In keeping with the labeling of the three "outer" threefold points of $\overline{\text{Ceva}}(1)$, let $n_{03}^{(q)}$ be the weight at the $(q + 2)$-fold point $x_1 = x_2 = 0$, let $n_{02}^{(q)}$ be the weight at the $(q + 2)$-fold point $x_1 = x_3 = 0$, and let $n_{01}^{(q)}$ be the weight at the $(q + 2)$-fold point $x_2 = x_3 = 0$. From equation (5.46) we have

$$\frac{1}{n_{14}^{(q)}} + \frac{1}{n_{24}^{(q)}} + \frac{q}{n_{12}^{(q)}} + \frac{2}{n_{03}^{(q)}} = q.$$

This is equivalent to

$$\frac{1}{n_{14}} + \frac{1}{n_{24}} + \frac{1}{n_{12}} + \frac{2}{q n_{03}^{(q)}} = 1.$$

Therefore, from the diophantine conditions (5.46) for $\overline{\text{Ceva}}(1)$, we must have $n_{03} = q n_{03}^{(q)}$, so that n_{03} is divisible by q. A similar computation holds for the other two $(q + 2)$-fold points, so that

$$n_{01}^{(q)} = \frac{n_{01}}{q}, \quad n_{02}^{(q)} = \frac{n_{02}}{q}, \quad n_{03}^{(q)} = \frac{n_{03}}{q},$$

and n_{01}, n_{02} are also divisible by q. For the q^2 triple points of $\overline{\text{Ceva}}(q)$, we simply recover, from (5.46), diophantine conditions equivalent to the remaining condition for $\overline{\text{Ceva}}(1)$, namely, $n_{04}^{(q)} = n_{04}$ and

$$\frac{1}{n_{12}} + \frac{1}{n_{13}} + \frac{1}{n_{23}} + \frac{2}{n_{04}} = 1.$$

Thus, the coverings of $\overline{\text{Ceva}}(1)$ with n_{14}, n_{24}, n_{34}, n_{01}, n_{02}, n_{03} divisible by q, and only those, come from $\overline{\text{Ceva}}(q)$. Now, in the passage from $\overline{\text{Ceva}}(1)$ to $\overline{\text{Ceva}}(q)$, we must notice two things. First, we assumed that all lines have weight at least 2 (see (†)), because, otherwise, we can ignore some of the lines of the arrangement. This can happen here, because $n_{14}^{(q)}$, $n_{24}^{(q)}$, or $n_{34}^{(q)}$ might be 1. In particular, if $n_{14}^{(q)}$, $n_{24}^{(q)}$, or $n_{34}^{(q)}$ are all 1, we recover Ceva(q). Secondly, by symmetry (see Chapter 7 and §5.7) any permutation of $\{1, 2, 3, 4\}$ could be applied, and, $n_{14}^{(q)}$, $n_{13}^{(q)}$, $n_{14}^{(q)}$, for example, may be taken as weights for the outside lines of $\overline{\text{Ceva}}(q)$. Making allowances for these possibilities, if we look at the fourteen solutions for Ceva(q) with $q \geq 3$ listed in Tables 5.6 to 5.10, we find that they correspond in this way, respectively, to the following cases of the list in Table 5.12: Nr $=$ 14, 19, 18, 15, 9, 5 ($q = 3$); Nr $=$ 21, 26, 2 ($q = 4$); Nr $=$ 17, 1 ($q = 5$); Nr $=$ 13, 22 ($q = 6$); Nr $=$ 3 ($q = 8$).

In the same way, we can also consider Ceva(2) as a cover of $\overline{\text{Ceva}}(1)$. As in Table 5.6 (for Ceva(3)), we have the possible solutions of (5.47) for Ceva(2), which also has threefold and twofolds points, given by the following table:

n	r	m
2	3	-4
3	3	∞
4	3	8
5	3	5
6	3	4
9	3	3

These are the six cases for Ceva(2) with equal weights for all lines. But Ceva(2) is equivalent to $\overline{\text{Ceva}}(1)$. In Table 5.12, the six cases are

$$\text{Nr} = 31, \quad 19, \quad 2, \quad 1, \quad 7, \quad 5.$$

As coverings of $\overline{\text{Ceva}}(1)$, they correspond to the cases

$$\text{Nr} = 31, \quad 25, \quad 4, \quad 6, \quad 12, \quad 16.$$

Recall that we can use permutations of $\{0, 1, \ldots, 4\}$. The cases Nr = 19 and 25; Nr = 2 and 4; Nr = 1 and 6; Nr = 7 and 12; and Nr = 5 and 16 are "commensurable," that is, they give rise to commensurable discontinuous groups acting on the two-dimensional complex ball.

So far, we have twenty cases of the list in Table 5.12 related to arrangements Ceva(q) where all lines have the same weight $n \geq 2$. Also, the exceptional cases Nr = 20 and 23 belong to Ceva(3) and Ceva(4), respectively, and have $n = \infty$. These have been excluded for the moment.

These twenty-two cases are exactly those where at least three of the μ_i are equal. The quintuples $(\mu_0, \mu_1, \ldots, \mu_4)$ giving rise to coverings of Ceva(q), $q \geq 2$, can be written in the form

$$\beta + 2\alpha, \quad \frac{1}{2} - \alpha, \quad \frac{1}{2} - \alpha, \quad \frac{1}{2} - \alpha, \quad \frac{1}{2} + \alpha - \beta, \qquad (5.49)$$

with $\alpha = \frac{1}{2n}$ and $\beta = \frac{1}{q}$. Indeed, μ_1, μ_2, μ_3 must be equal, and x_{14}, x_{24}, x_{34} must vanish (see (5.48)); for example,

$$x_{14} = 1 - q + q \left(\frac{1}{2} - \alpha + \frac{1}{2} + \alpha - \beta \right) = 1 - \beta q$$

vanishes for $\beta = \frac{1}{q}$.

We can also consider the coverings of $\overline{\text{Ceva}}(q)$, $q \geq 1$, with equal weights for all lines. Since $\overline{\text{Ceva}}(q)$ has singular points of order 3 and $q + 2$, exactly those n are possible that work for Ceva($q + 2$) and that are listed in Tables 5.6 to 5.10. The quintuples for $\overline{\text{Ceva}}(1)$ listed in Table 5.12 corresponding to these coverings can be written in the form (5.49), with $\alpha = \frac{1}{2n}$ and $\beta = \frac{1}{nq}$. For $q = 1$, we have $n = 2, 3, 4, 5, 6, 9$, from Table 5.6, corresponding to the cases in Table 5.13 given by

$$\text{Nr} = 31, \quad 19, \quad 2, \quad 1, \quad 7, \quad 5.$$

Here, (5.49) has the form $\left(\frac{4}{2n}, \frac{n-1}{2n}, \frac{n-1}{2n}, \frac{n-1}{2n}, \frac{n-1}{2n} \right)$. Furthermore, the cases $q = 2$ and $n = 2, 3, 4$, correspond to the cases in Table 5.13 given by

$$\text{Nr} = 21, \quad 22, \quad 2.$$

Moreover, the cases $q = 3$ and $n = 2, 5$ correspond to the cases

$$\text{Nr} = 13, \quad 15,$$

and the cases $q = 4$ and $n = 2, 3$ correspond to the cases

$$\text{Nr} = 3, \quad 26.$$

Finally, $q = 6$ and $n = 2$ correspond to the case

$$\text{Nr} = 13.$$

5.6 BLOWING DOWN RATIONAL CURVES AND REMOVING ELLIPTIC CURVES

In §5.4 we allowed some of the $\widetilde{E}_j$ lying over the blown-up points E_j to be a union of rational or elliptic curves. The rational case corresponded to a weight $m_j < 0$ for E_j in (5.46) and the rational curves $\widetilde{E}_j$ over E_j were exceptional curves with self-intersection -1 which were then blown down. The elliptic case corresponded to a weight $m_j = \infty$ for E_j in (5.46) and the elliptic curves $\widetilde{E}_j$ over E_j were removed. We now study this situation in more generality, allowing some of the $\widetilde{D}_i$ lying over the proper transforms of the lines of the arrangement to be a union of rational or elliptic curves. We are therefore no longer assuming the conditions $n_i \geq 2$ and n_i finite of (†) in §5.5. If X denotes the blow-up of $\mathbb{P}_2$ at the singular points of a line arrangement, we are focusing on finite covers of blow-downs of X that are ball quotients rather than finite covers of blow-ups of the arrangement that are ball quotients. Suppose that, for the indices marked k and the indices marked ℓ, all the curves E_ℓ and D_k are pairwise disjoint, and all the curves $\widetilde{E}_\ell$ and $\widetilde{D}_k$ are either rational or elliptic.

We always assume that the E_ℓ and D_k, with $\widetilde{E}_\ell$ rational or elliptic and with $\widetilde{D}_k$ rational or elliptic, are mutually disjoint on X.

The corresponding numbers P_ℓ and Q_k given by (5.14) and (5.15) are therefore negative, since in §5.1 we assumed $n_k > 0$ for D_k and $m_\ell > 0$ for E_ℓ. This is because, in §5.1, the n_k and m_ℓ corresponded to ramification indices rather than weights. To avoid confusion with the conventions of §5.5, where m_ℓ denotes a weight that could be negative or infinite, we anticipate the notation of Chapter 6, Theorem 6.4, and assign $n'_k > 0$ to D_k and $m'_\ell > 0$ to E_ℓ as the ramification indices of §5.1. For the D_i and E_j not corresponding to the rational or elliptic case, we also write $n'_i > 0$ and $m'_j > 0$ for the ramification indices. With our new notation, the formulae (5.14) and (5.15) become

$$P_\ell = -\frac{2}{m'_\ell} + r_\ell - 2 - \sum_{i \sim \ell} \frac{1}{n'_i}$$

and

$$Q_k = \frac{-2(\sigma_k - 1)}{n'_k} + \tau_k - 2 - \sum_{j \sim k} \frac{1}{m'_j} - \sum_{i \sim k} \frac{1}{n'_i}.$$

In the rational case, the numbers P_ℓ and Q_k will vanish if we replace m'_ℓ in P_ℓ by a uniquely determined weight $m_\ell < 0$, and n'_k in Q_k by a uniquely determined weight $n_k < 0$. For P_ℓ, this is equivalent to (5.46) with $m_\ell < 0$.

We now choose the ramification indices by setting $n'_k = -n_k$ for D_k and $m'_\ell = -m_\ell$ for E_ℓ. In the elliptic case, the P_ℓ or Q_k can be made to vanish by replacing m'_ℓ and n'_k by the weight ∞. For P_ℓ, this is equivalent to (5.46) with $m_\ell = \infty$. In the elliptic case, we choose arbitrary positive numbers n'_k for D_k and m'_ℓ for E_ℓ as ramification indices.

In both the rational and elliptic cases, the weights m_ℓ and n_k are given by the diophantine conditions

$$\frac{2}{m_\ell} = r_\ell - 2 - \sum_{i \sim \ell} \frac{1}{n'_i},$$

$$\frac{2(\sigma_k - 1)}{n_k} = \tau_k - 2 - \sum_{j \sim k} \frac{1}{m'_j} - \sum_{i \sim k} \frac{1}{n'_i},$$

which we shall encounter again in Chapter 6 in formula (6.3).

First, we study the rational case. We have

$$P_\ell = -\frac{2}{m'_\ell} + \frac{2}{m_\ell} = -\frac{4}{m'_\ell}$$

and

$$Q_k = -\frac{2(\sigma_k - 1)}{n'_k} + \frac{2(\sigma_k - 1)}{n_k} = -\frac{4(\sigma_k - 1)}{n'_k}.$$

Therefore, arguing as in §5.4, and using Definition 5.2 of §5.2, as well as the equations for P_ℓ and Q_k that follow it, we have

$$\text{prop}(\widetilde{E_\ell}) = -\frac{4N}{(m'_\ell)^2}, \qquad \text{prop}(\widetilde{D_k}) = -\frac{4N(\sigma_k - 1)}{(n'_k)^2},$$

$$\widetilde{E_\ell} \cdot \widetilde{E_\ell} = -\frac{N}{(m'_\ell)^2}, \qquad \widetilde{D_k} \cdot \widetilde{D_k} = -\frac{N(\sigma_k - 1)}{(n'_k)^2},$$

$$e(\widetilde{E_\ell}) = \frac{2N}{(m'_\ell)^2}, \qquad e(\widetilde{D_k}) = \frac{2N(\sigma_k - 1)}{(n'_k)^2}.$$

It follows that $\widetilde{E_\ell}$ is a disjoint union of $\frac{N}{(m'_\ell)^2}$ exceptional curves, as observed earlier in §5.4, and $\widetilde{D_k}$ is a union of $\frac{N(\sigma_k-1)}{(n'_k)^2}$ exceptional curves. When we blow down these exceptional curves, which we have assumed are disjoint, this affects the curves $\widetilde{D_i}$ with $i \sim \ell$ or $i \sim k$. Namely, by formula (3.29) of Chapter 3, the number $\text{prop}(\widetilde{D_i})$ goes up by $2\widetilde{E_\ell} \cdot \widetilde{D_i}$ and $2\widetilde{D_k} \cdot \widetilde{D_i}$ for each rational $\widetilde{E_\ell}$ and $\widetilde{D_k}$. Since

$$2\widetilde{E_\ell} \cdot \widetilde{D_i} = \frac{2N}{m'_\ell n'_i}, \qquad 2\widetilde{D_k} \cdot \widetilde{D_i} = \frac{2N}{n'_k n'_i},$$

and $n'_i > 0$ by assumption, this can be compensated for by replacing, in the formula (see (5.15)) for Q_i, the term $-\frac{1}{m'_\ell} = \frac{1}{m_\ell}$ or $-\frac{1}{n'_k} = \frac{1}{n_k}$, respectively, by $\frac{1}{m'_\ell} = -\frac{1}{m_\ell}$ or $\frac{1}{n'_k} = -\frac{1}{n_k}$, since

$$\text{prop}(\widetilde{D_i}) = \frac{N}{n'_i} Q_i.$$

The same procedure can be applied to the curves $\widetilde{E_j}$ with $j \sim k$ for some exceptional $\widetilde{D_k}$. Hence, after the blow-downs, the prop of the transforms of $\widetilde{D_i}, \widetilde{E_j}$ can be written as in (5.14) or (5.15), but now using the negative integers $-m'_k = m_k, -n'_\ell = n_\ell$ instead of m'_ℓ, n'_k for the rational curves $\widetilde{E_\ell}, \widetilde{D_k}$ that are blown down. The value of $\text{Prop}(Y)$ goes down by the sum of all the $\frac{4N}{(m'_\ell)^2}$ and $\frac{4N(\sigma_k-1)}{(n'_k)^2}$ taken over the blow-downs of the rational curves $\widetilde{E_\ell}, \widetilde{D_k}$.

In the elliptic case, we remove from the finite cover Y the union, $\widetilde{C}$, of all elliptic $\widetilde{E_\ell}$ and $\widetilde{D_k}$. As in §5.4, we have

$$\text{Prop}(Y, \widetilde{C}) = \text{Prop}(Y) + \widetilde{C} \cdot \widetilde{C}.$$

Therefore, $\text{Prop}(Y, \widetilde{C})$ is the same as $\text{Prop}(Y)$ with $-\widetilde{C} \cdot \widetilde{C}$ subtracted. As

$$\widetilde{E}_\ell \cdot \widetilde{E}_\ell = -\frac{N}{(m'_\ell)^2}, \qquad \widetilde{D}_k \cdot \widetilde{D}_k = -\frac{N(\sigma_k - 1)}{(n'_k)^2},$$

the quantity $-\widetilde{C} \cdot \widetilde{C}$ is the sum of all the $\frac{N}{(m'_\ell)^2}$, $\frac{N(\sigma_k-1)}{(n'_k)^2}$ taken over the elliptic curves $\widetilde{E}_\ell$, $\widetilde{D}_k$.

Let Y' be the smooth compact surface obtained from Y by blowing down the union of disjoint rational curves $\widetilde{D}_k$, $n_k < 0$, and $\widetilde{E}_\ell$, $m_\ell < 0$. Then, Y' is a smooth compact surface and, for any smooth curve $\widetilde{D}$ on Y', we have, by Definition 5.2 of §5.2,

$$\text{prop}(\widetilde{D}) = 2\widetilde{D} \cdot \widetilde{D} - e(\widetilde{D}) = 2c_1(Y) \cdot \widetilde{D} - 3e(\widetilde{D}),$$

using the adjunction formula. Let C be the union of the disjoint elliptic curves given by the image in Y' of the union $\widetilde{C}$ of the $\widetilde{D}_k$, $n_k = \infty$, and $\widetilde{E}_\ell$, $m_\ell = \infty$. For the non-compact surface, $Y' \setminus C$, as we explained in §5.4, we replace $c_1(Y)$ by

$$c_1(Y, C) = c_1(Y) - C$$

and define, for any smooth curve $\widetilde{D}$ on Y' that intersects C transversally,

$$\text{prop}(\widetilde{D}, C) = 2(c_1(Y, C)) \cdot \widetilde{D} - 3e(\widetilde{D} \setminus \widetilde{D} \cap C) = \text{prop}(\widetilde{D}) + \widetilde{D} \cdot C.$$

Removing C from Y', we replace every $\text{prop}(\widetilde{E}_j)$ by $\text{prop}(\widetilde{E}_j, C)$, and every $\text{prop}(\widetilde{D}_i)$ by $\text{prop}(\widetilde{D}_i, C)$. This corresponds to taking (5.14) and (5.15) (with $n_k = n'_k$ and $m_\ell = m'_\ell$) and putting ∞ in place of n'_k for $\widetilde{D}_k$ elliptic, and ∞ in place of m'_ℓ for $\widetilde{E}_\ell$ elliptic. We can now generalize Höfer's formula (5.18):

Theorem 5.4 *Consider a blown-up line arrangement X with weights n_i for the proper transforms D_i of the lines, and m_j for the blown-up singular points E_j, where n_i and m_j are integers or ∞. Let D_k, E_ℓ be the curves whose associated weights are negative or ∞ and suppose that these curves are pairwise disjoint. Consider a good covering Y of X with the given weights n_i, m_j as ramification indices, except at the D_k, E_ℓ, where the weights are to be determined by*

$$\frac{2(\sigma_k - 1)}{n_k} = \tau_k - 2 - \sum_{j \sim k} \frac{1}{m_j} - \sum_{i \sim k} \frac{1}{n_i}, \qquad \frac{2}{m_\ell} = r_\ell - 2 - \sum_{i \sim \ell} \frac{1}{n_i}$$

and the ramification indices are $-n_k$, $-m_\ell$ if n_k, m_ℓ are negative, or are arbitrary if n_k, m_ℓ are ∞. If a weight is negative, the corresponding curves in Y are exceptional rational curves and can be blown down to give a smooth surface Y'. If a weight is ∞, the corresponding curves in Y and Y' are elliptic. Let C be the union of all the elliptic curves in Y'. Then,

$$\text{Prop}(Y', C) = \frac{1}{2}\sum_i (n_i - 1)\text{prop}(\widetilde{D_i}, C) - \frac{1}{2}\sum_j (m_j + 1)\text{prop}(\widetilde{E_j}, C).$$

(5.50)

The sum is taken only over those i, j with positive weights n_i, m_j that are not ∞. This formula applies to the non-compact surface $Y' \setminus C$. The $\text{prop}(\widetilde{D_i}, C)$ refer to the transforms on Y' of the curves $\widetilde{D_i}$, and we may write $\text{prop}(\widetilde{D_i})$ if $\widetilde{D_i}$ does not intersect C. The same holds for $\text{prop}(\widetilde{E_j}, C)$.

Proof. Recall Höfer's formula of §5.2, and that we rewrote it there as formula (5.19) for $\text{Prop}(Y)$ in terms of the ramification indices. In the notation of the discussion of this section, (5.19) becomes

$$\text{Prop}(Y) = \frac{1}{2}\left(\sum_i (n'_i - 1)\text{prop}(\widetilde{D_i}) - \sum_j (m'_j + 1)\text{prop}(\widetilde{E_j}) \right)$$

$$+ \frac{1}{2}\left(\sum_k (n'_k - 1)\text{prop}(\widetilde{D_k}) - \sum_\ell (m'_\ell + 1)\text{prop}(\widetilde{E_\ell}) \right).$$

The first sum is taken over those i, j with positive weights $n_i = n'_i$, $m_j = m'_j$ that are not ∞. The second sum is taken over the k, ℓ with weights n_k, m_ℓ negative or infinite.

In the rational case, that is, when n_k, m_ℓ are negative, we have $n'_k = -n_k$, $m'_\ell = -m_\ell$. Consider the effect of blowing down $\overline{E}_\ell$, $m_\ell < 0$. From the discussion preceding the theorem, the quantity $\text{Prop}(Y)$ goes down by $\frac{4N}{m_\ell^2}$, whereas

$$-\frac{1}{2}(|m_\ell| + 1)\text{prop}(\widetilde{E_\ell}) = (|m_\ell| + 1)\frac{2N}{m_\ell^2}$$

is omitted on the right-hand side of Höfer's formula (5.19). The number $\frac{1}{2}(n_i - 1)\text{prop}(\widetilde{D_i}) = \frac{1}{2}(n_i - 1)\frac{N}{n_i}Q_i$ with $i \sim \ell$ is increased by $\frac{1}{2}(n_i - 1)\frac{2N}{n_i|m_\ell|}$ by Lemma 5.1 of §5.2. Using the diophantine condition

$$\frac{2}{m_\ell} = r_\ell - 2 - \sum_{i \sim \ell} \frac{1}{n_i},$$

we deduce that the sum of these increases is

$$\left(-\sum_{i\sim\ell}\frac{1}{n_i}+r_\ell\right)\frac{N}{|m_\ell|}=\left(-\frac{2}{|m_\ell|}+2\right)\frac{N}{|m_\ell|}.$$

It follows that, after the blow-down of $\widetilde{E_\ell}$, $m_\ell < 0$, the right-hand side of Höfer's formula (5.19) goes down by

$$(|m_\ell|+1)\frac{2N}{m_\ell^2}+\left(\frac{2}{|m_\ell|}-2\right)\frac{N}{|m_\ell|}=\frac{4N}{m_\ell^2},$$

which balances the decrease in the left-hand side Prop(Y). Therefore, Höfer's formula remains valid under the blow-downs of the rational $\widetilde{E_\ell}$. The effect of blowing down $\widetilde{D_k}$, $n_k < 0$, can be computed in an analogous way, although the exact details are a little more complicated. We leave this as an exercise for the reader.

Now consider the effect of removing the elliptic $\widetilde{E_\ell}$ from Y'. The quantity Prop(Y') goes down by $\frac{N}{m_\ell'^2}$, as remarked before the theorem. The term

$$-\frac{1}{2}(m_\ell'+1)\text{prop}(\widetilde{E_\ell})=(m_\ell'+1)\frac{N}{m_\ell'^2}=\frac{N}{m_\ell'^2}+\frac{N}{m_\ell'}$$

is omitted. The number $\frac{1}{2}(n_i-1)\text{prop}(\widetilde{D_i})=\frac{1}{2}(n_i-1)\frac{N}{n_i}Q_i$ with $i\sim\ell$ is increased by $\frac{1}{2}(n_i-1)\frac{N}{n_i m_\ell'}$. Using the diophantine condition

$$r_\ell-2-\sum_{i\sim\ell}\frac{1}{n_i}=0,$$

it follows that the sum of these increases is

$$\frac{N}{2m_\ell'}\left(-\sum_{i\sim\ell}\frac{1}{n_i}+r_\ell\right)=\frac{N}{m_\ell'}.$$

Therefore, overall the right-hand side of Höfer's formula decreases by $\frac{N}{m_\ell'^2}$, which balances the decrease in Prop(Y'). Therefore, Höfer's formula remains valid after removing the elliptic curves $\widetilde{E_\ell}$ from Y'. We can compute in an analogous way the effect of removing the $\widetilde{D_k}$ with $n_k=\infty$. We also leave this as an exercise for the reader. ☐

The arguments in the preceding proof show that Höfer's formula in the theorem holds if we blow down only some of the rational curves and remove only some of the elliptic curves. Therefore, in the situation of §5.4, we can start with Y, fulfill $P_j = 0$ for all the singular points of the arrangement allowing solutions with $m_j > 0$, with $m_j < 0$, and with $m_j = \infty$, blow down the rational $\widetilde{E}_j$ with $m_j < 0$ on Y to get Y', and remove the image C of the $\widetilde{E}_j$ with $m_j = \infty$ to get $Y' \setminus C$. We then have

$$\mathrm{Prop}(Y', C) = \frac{1}{2} \sum_i (n_i - 1)\mathrm{prop}(\widetilde{D}_i, C). \tag{5.51}$$

On the other hand, by (5.45),

$$\mathrm{Prop}(Y', C) = \frac{N}{4} \sum_{i,l} R_{il} x_i x_l. \tag{5.52}$$

Comparing (5.51) and (5.52) term by term, we conclude that

$$(n_i - 1)\mathrm{prop}(\widetilde{D}_i, C) = \frac{N}{2} x_i \sum_l R_{il} x_l. \tag{5.53}$$

If we now blow down the rational $\widetilde{D}_i$ and remove the images of the elliptic $\widetilde{D}_i$, then, with $x_i = 1 - \frac{1}{n_i}$, and $n_i < 0$ or $n_i = \infty$, the $\mathrm{prop}(\widetilde{D}_i, C)$ vanish formally and, for other i, formula (5.53) gives the $\mathrm{prop}(\widetilde{D}_i, C)$ in terms of the matrix R. Therefore, the vanishing of all prop numbers in (5.50) is equivalent to the vanishing of all P_j in (5.37), together with all $(x_1, \ldots, x_k)$ belonging to the kernel of R. Here, negative weights, and ∞ as a weight, are allowed provided the lines with these weights are disjoint.

In the examples of §5.5, the matrix R was always positive semi-definite. The vanishing of the quadratic form $\sum_{il} R_{il} x_i x_l$ is therefore equivalent to $(x_1, \ldots, x_k)$ belonging to the kernel of R.

Extending the possibilities for the assigned weights

We now extend the list of possibilities for the assigned weights on the line arrangements blown up at their singular points, using the results of this section. We can now include the exceptional cases corresponding to Nrs 20 and 23 in Table 5.12, which belong, respectively, to Ceva(3) and Ceva(4) with $n = \infty$. For the complete quadrilateral, all twenty-seven cases have now been treated.

Next, we consider the Hesse, icosahedral, G_{168}, A_6, and extended Hesse arrangements.

For the G_{168} configuration, we only have the additional case $n = \infty$. Moreover, all twenty-one lines $\widetilde{D}_i$ are disjoint because there are no regular points. For the triple and quadruple points, m is 2 and 1, respectively, and (5.47) is satisfied.

For the extended Hesse configuration, let $m_{(4)}$ be the weight for the blown-up quadruple points, and $m_{(5)}$ the weight for the blown-up quintuple points. The equations

$$\frac{4}{n_H} + \frac{2}{m_{(4)}} = 2,$$

$$\frac{2}{n_H} + \frac{3}{n_C} + \frac{2}{m_{(5)}} = 3$$

are satisfied for the values in the following table:

n_H	n_C	$m_{(4)}$	$m_{(5)}$
2	∞	∞	1
∞	3	1	1

Otherwise, there are no more solutions for these arrangements.

5.7 TABLES OF THE WEIGHTS GIVING PROP = 0

We end this chapter by returning to the list of tables that summarize the solutions to $\mathrm{Prop}(Y', C) = 0$ given by the weighted line arrangements of §5.3. Therefore, we require (5.46) of §5.5 to be true for all m_j and the vector $(x_1, x_2, \ldots, x_k)$, where $x_i = 1 - \frac{1}{n_i}$ and k is the number of lines of the arrangement, to be in the kernel of the matrix R of §5.2. As in §5.3, the line arrangements we study are all those known to satisfy $3\tau_i = k + 3$, where τ_i is the number of intersection points on the ith line and $k > 3$. Recall that the n_i are the weights on the proper transforms of the lines of the arrangement and that the m_j are the weights on the exceptional curves given by blow-ups of the singular points of the arrangement. We also assume (†) of §5.5, namely,

(†) We allow only $n_i \geq 2$ and finite, but permit $m_j \geq 1$, $m_j < 0$, or $m_j = \infty$.

In §5.6 we give the extra solutions to $\mathrm{Prop}(Y', C) = 0$ that can occur when we also allow the n_i to be negative or infinite.

We repeat and summarize here some of the discussion in §§5.5 and 5.6 in order to help the reader to follow these tables. First, Table 5.1 gives all the solutions to (5.47) assuming (†) and $r \geq 3$. Then, for each arrangement having corank$(R)=1$, so that the lines have equal weight n, Tables 5.2 through 5.10 give the subset of solutions listed in Table 5.1 that are possible for that arrangement (see §5.5). To recap, the assumptions in these tables amount to the following.

Recall that r is the number of lines of weight n, assuming $n \geq 2$, passing through a singular point of the arrangement (hence $r \geq 3$). This singular r-fold point is blown up to an exceptional divisor E which is assigned the weight m. The weight m is allowed to take any integer value, positive or negative, and is allowed to be infinite, although it must of course satisfy the diophantine condition (5.46). For the extended Hesse arrangement, we have corank(R)=2, and the arrangement consists of the lines from the Hesse arrangement, which must have equal weight n_H, and those of the Ceva(1) arrangement, which must have equal weights n_C. This arrangement has r-fold points, where $r \geq 3$, with $r = 4, 5$. This yields two diophantine conditions derived from (5.46), which are given and solved in §5.5, with the results listed in the two tables of that section. The condition at the fourfold points only involves n_H, and at the fivefold points involves n_H and n_C. The table of solutions of the diophantine condition at the fivefold points gives possibilities for n_H that are exactly the same as those possibilities for n_H at the fourfold points. The possibilities for (n_H, n_C) given by Table 5.11 are therefore taken from the table for the fivefold points in §5.5. For each choice of (n_H, n_C), we have a corresponding unique value m_4 of m at the fourfold points, and a unique value m_5 of m at the fivefold points, which we also give in the table. Again, the weight m is assigned to the exceptional curve given by the blow-up of these multiple intersection points.

By far the most complex case is the extended Ceva arrangement $\overline{\text{Ceva}}(q)$, $q \geq 1$, which has corank(R) = 4. In §5.5, we began by considering $\overline{\text{Ceva}}(1)$, which is the complete quadrilateral also discussed in Chapter 3, §3.3. The arrangement has six lines, three two-fold points, and four three-fold points. We refer the reader to §§3.3 and 5.5 for notation. In particular, let X be the surface obtained from $\overline{\text{Ceva}}(1)$ by blowing up the four three-fold points. This surface has ten divisors, $D_{\alpha\beta} = D_{\beta\alpha}$, where $\alpha, \beta = 0, 1, 2, 3, 4$, as follows. The $D_{\alpha\beta}$ with $\alpha, \beta = 1, 2, 3, 4$, are the proper transforms of the lines of $\overline{\text{Ceva}}(1)$, and the $D_{0\beta}$, $\beta = 1, 2, 3, 4$, are the exceptional curves corresponding to the blown-up triple points. Recall that the kernel of R is parametrized by four numbers, $\mu_1, \mu_2, \mu_3, \mu_4$, and that we introduced a fifth number given by $\mu_0 = 2 - \mu_1 - \mu_2 - \mu_3 - \mu_4$. We define $n_{\alpha\beta} = n_{\beta\alpha} = (1 - \mu_\alpha - \mu_\beta)^{-1}$, which is the weight assigned to $D_{\alpha\beta}$. For $\alpha, \beta = 1, 2, 3, 4$, we assume in §5.5 that the $n_{\alpha\beta} = n_{\beta\alpha}$ are integers at least 2, in order to satisfy (†). In §5.5, we discussed the list of Le Vavasseur, who found all quintuples $(\mu_i)_{i=0}^5$, with μ_i rational, such that the $n_{\alpha\beta}$ are nonzero integers, positive or negative, or are infinity. This list is given in Table 5.13, with the notation $\mu_i = \alpha_i/d$, $d \geq 1$. The column "DM" refers to Deligne-Mostow's list [33] of weights giving rise to finite covers that are ball quotients, as explained in §5.5. In Table 5.12, we rewrite entries 1 through 27 and 31 in terms of the weights $n_{\alpha\beta}$: the other cases from Le Vavasseur's list are excluded as the weight 1 occurs for at least one of the $D_{\alpha\beta}$. Strictly speaking, under the assumptions of §5.5, the cases 20 and 23 in Table 5.12 have to be omitted because (†) is not satisfied, whereas for the others it is satisfied. We incorporate these cases in the later section, §5.6. Because of the symmetry of the blown-up complete quadrilateral X, we can in any case permute the weights, so these cases can be incorporated that way as well, there being permutations of the weights of 20 and 23 that do satisfy (†). As we remark in Section 5.5, the cases 1 through 27 lead to ball quotients, whereas, although 31 satisfies (†), it leads to an abelian surface, which is why it is not in the Deligne-Mostow list. The symmetry of X referred to above is discussed fully in Chapter 7. Let S_5 act as permutations of $\{0, 1, 2, 3, 4\}$. Then, there is an induced action on the subscripts $(\alpha\beta)$ of the ten divisors $D_{\alpha\beta} = D_{\beta\alpha}$, $\alpha, \beta = 0, 1, 2, 3, 4$. We show in Chapter 7 that the induced permutation of the $D_{\alpha\beta} = D_{\beta\alpha}$ preserves their pairwise intersection numbers and extends to an action on X. Therefore, taking into account our remarks about Le Vavasseur's list in Table 5.13, for every entry in Table 5.12, and every $\sigma \in S_5$, we may assign the weight $n_{\sigma^{-1}(\alpha)\sigma^{-1}(\beta)} = n_{\sigma^{-1}(\beta)\sigma^{-1}(\alpha)}$ to $D_{\alpha\beta} = D_{\beta\alpha}$ and get allowed weights for $\overline{\text{Ceva}}(1)$, none of which equals 1. Moreover, the weights given in Table 5.12, as well as all their permutations, give the full set of solutions to (5.46), satisfying (†) or not, with none of the $n_{\alpha\beta}$ equal to 1.

In §5.5, we then turn to $\overline{\text{Ceva}}(q)$, for $q > 1$, and use the fact that $\overline{\text{Ceva}}(1)$ lifts to $\overline{\text{Ceva}}(q)$ under the map $[x_0 : x_1 : x_2] \to [x_0^q : x_1^q : x_2^q]$ of $\mathbb{P}_2$ to itself. The arrangement $\overline{\text{Ceva}}(q), q > 1$, has $3q + 3$ lines, $3q$ double points, q^2 triple points, and three $(q + 2)$-fold points. For the weights on $\overline{\text{Ceva}}(q)$, let $n_{14}^{(q)}$ be the weight associated to the line $x_1 = 0$, let $n_{24}^{(q)}$ be the weight associated to $x_2 = 0$, and let $n_{34}^{(q)}$ be the weight associated to $x_3 = 0$. Let $n_{12}^{(q)}$ be the weight associated to the q lines $x_1^q = x_2^q$, let $n_{23}^{(q)}$ be the weight associated to the q lines $x_1^q = x_3^q$, and let $n_{13}^{(q)}$ be the weight associated to the q lines $x_1^q = x_3^q$. Let $n_{01}^{(q)}$ be the weight at the $(q + 2)$-fold point $x_2 = x_3 = 0$, let $n_{02}^{(q)}$ be the weight at the $(q + 2)$-fold point $x_1 = x_3 = 0$, and let $n_{03}^{(q)}$ be the weight at the $(q + 2)$-fold point $x_1 = x_2 = 0$. Finally, let $n_{04}^{(q)}$ be the weight at the q^2 triple points. Let $(\alpha, \beta, \gamma, \delta, \varepsilon)$ be the image of $(0, 1, 2, 3, 4)$ under a permutation in S_5. Consider the following equations, which, by the arguments of §5.5, give the weights of $\overline{\text{Ceva}}(q)$ satisfying (5.46) in terms of the (non-permuted) original entries in any row of Table 5.12:

$$n_{14}^{(q)} = \frac{n_{\beta\varepsilon}}{q}, \; n_{24}^{(q)} = \frac{n_{\gamma\varepsilon}}{q}, \; n_{34}^{(q)} = \frac{n_{\delta\varepsilon}}{q},$$

$$n_{12}^{(q)} = n_{\beta\gamma}, \; n_{23}^{(q)} = n_{\gamma\delta}, \; n_{13}^{(q)} = n_{\beta\delta},$$

$$n_{01}^{(q)} = \frac{n_{\alpha\beta}}{q}, \; n_{02}^{(q)} = \frac{n_{\alpha\gamma}}{q}, \; n_{03}^{(q)} = \frac{n_{\alpha\delta}}{q}, \; n_{04}^{(q)} = n_{\alpha\delta}.$$

In order for these weights to satisfy (†), we must have $n_{\beta\varepsilon}$, $n_{\gamma\varepsilon}$, $n_{\delta\varepsilon}$ positive, divisible by q, and greater than q; we must have $n_{\beta\gamma}$, $n_{\gamma\delta}$, $n_{\beta\delta}$ finite positive integers at least 2; and, finally, we must have $n_{\alpha\beta}$, $n_{\alpha\gamma}$, $n_{\alpha\delta}$ divisible by q. There is no condition on the weight $n_{\alpha\delta}$. For example, Nr=1 of Table 5.12 does not give rise to weights satisfying (5.46) and (†) for any $\overline{\text{Ceva}}(q)$, $q > 1$. On the other hand, any permutation of the weights of Nr=2 of Table 5.12 gives such weights for $\overline{\text{Ceva}}(2)$ since the entries of the row are all positive and divisible by $2 \times 2 = 4$. Notice that Nr=2 with $(\alpha, \beta, \gamma, \delta, \varepsilon) = (0, 1, 2, 3, 4)$ does not give weights for $\overline{\text{Ceva}}(4)$ satisfying (†) since the lines $x_1 = 0$, $x_2 = 0$, $x_3 = 0$ have weight 1 (giving instead weights for Ceva(4)). However, Nr=2 with $(\alpha, \beta, \gamma, \delta, \varepsilon) = (4, 1, 2, 3, 0)$ gives

$$n_{14}^{(q)} = \frac{n_{10}}{4} = \frac{n_{01}}{4} = 2, \; n_{24}^{(q)} = \frac{n_{20}}{4} = \frac{n_{02}}{4} = 2, \; n_{34}^{(q)} = \frac{n_{30}}{4} = \frac{n_{03}}{4} = 2,$$

$$n_{12}^{(q)} = n_{12} = 4, \; n_{23}^{(q)} = n_{23} = 4, \; n_{13}^{(q)} = n_{13} = 4,$$

$$n_{01}^{(q)} = \frac{n_{41}}{4} = \frac{n_{14}}{4} = 1, \; n_{02}^{(q)} = \frac{n_{42}}{4} = \frac{n_{24}}{4} = 1, \; n_{03}^{(q)} = \frac{n_{43}}{4} = \frac{n_{34}}{4} = 1,$$

$$n_{04}^{(q)} = n_{40} = n_{04} = 8,$$

which satisfies (†), the weights equal to 1 being now at the three $(q + 2)$-fold points. To list all the weights for $\overline{\text{Ceva}}(q)$, $q > 1$, in terms of the entries of Table 5.12 and their permutations is therefore a lengthy process, and we content ourselves here with providing the above recipe for compiling this list. In §5.5, we list all rows of Table 5.12 that give rise to equal weights on all the lines of $\overline{\text{Ceva}}(q)$, $q > 1$.

Chapter Six

Existence of Ball Quotients Covering Line Arrangements

In Chapter 5, we assumed the existence of certain finite covers of weighted line arrangements in the projective plane and studied necessary conditions for these covers to be ball quotients. In this chapter, we justify this existence assumption. Our main references are [83], [84], and the references given in these papers. Other important references will be cited where relevant. In particular, we follow closely the sequence of papers by R. Kobayashi, [79], [80], [81], [82]. We aim to give an introduction to these papers for the nonspecialist in the special situation of the weighted line arrangements of Chapter 5, although we do quote the more general results of [83] in §6.1. Similar results to those of [83], [84] are due independently to S. Y. Cheng-S. T. Yau [27]. The global finite orbifold uniformization problem of weighted line arrangements in the projective plane was also studied by Kato [74], [75].

We do not treat in full all the ingredients needed to prove the main result, namely Theorem 6.4 of this chapter, since this is beyond the scope of this book. For example, we use without proof deep results from the theory of partial differential equations similar to those used in Chapter 4, §4.3, where we discussed the Miyaoka-Yau inequality for surfaces. We do, however, present a self-contained discussion with complete references, which should suffice for the interested reader who wants to reconstruct a fully detailed treatment. To make the comparison between our exposition and those of [83], [84] clear, we have tried as much as possible to explain both the notation of those references and ours.

We begin by giving a qualitative outline of the discussion of this chapter. The main goal is to proceed in a fashion similar to that of Chapter 4, §4.3, but with canonical divisors (called *log-canonical*) and Euler numbers reflecting the weight data on divisors on the blow-up X of $\mathbb{P}_2$ at the singular points of a line arrangement. This weight data is encoded by a $\mathbb{Q}$-divisor D on X. To (X, D) we associate the so-called *log-canonical model* (X'', D'') obtained by blowing down the divisors on X with negative weight and contracting the divisors with infinite weight. The resulting singularities on (X'', D'') are all quotient singularities whose local uniformizations carry a Kähler–Einstein metric. We remove from (X'', D'') the points given by the images in X'' of the divisors of infinite weight to form the orbifold (X_0'', D_0''). We then prove a Miyaoka–Yau inequality for (X_0'', D_0'') that is expressed in terms of the self-intersection of the *log-canonical divisor* $K_{X''} + D''$ and a suitable Euler number. The key ingredient, as in Chapter 4, §4.3, is the existence of a global Kähler–Einstein metric, which is now an orbifold metric on (X_0'', D_0''), whose associated form is in the same cohomology class as that of a given Kähler metric. That Kähler metric is constructed by patching together the Kähler–Einstein metrics on the local uniformizations of the singularities of (X_0'', D_0'') with a Kähler metric induced on X'' via a projective embedding given by the log-canonical divisor $K_{X''} + D''$, which we have to show is ample. The existence of the required Kähler–Einstein metric relies on deep results on partial differential equations. We then use the Kähler–Einstein property to prove an inequality between Chern forms that, when integrated, gives the appropriate Miyaoka–Yau inequality. In the case of equality, curvature considerations and the construction of the original Kähler metric ensure that (X_0'', D_0'') is an orbifold ball quotient $\Gamma \backslash B_2$ and that the ramification of the map from B_2 to X_0'' matches the weight data on D_0''. For any normal subgroup Γ' of Γ acting without fixed points, the free ball quotient $\Gamma' \backslash B_2$ gives a sought-after ramified cover of X_0'', as described in Theorem 5.4 of Chapter 5 and Theorem 6.4 of this chapter.

In order to relate our discussion to the key references [83] and [84], we will carefully reconsider the canonical divisors and Euler numbers defined in Chapter 5 on the finite covers of weighted line arrangements. As we wish to show that ball quotient finite covers exist, we will instead work with the X and X'' described above and with divisors on these surfaces with coefficients in $\mathbb{Q}$ (called $\mathbb{Q}$-divisors). The relevant canonical divisors and Euler numbers on these surfaces will then be constructed so as to lift to their correct counterparts on the finite covers whose existence we shall establish.

In what follows, we will often encounter the addition "log-" to various words. This terminology has its origin in the consideration of varieties with differential 1-forms having logarithmic singularities along a given divisor. Let D be a divisor, whose irreducible components all meet transversally, on a smooth compact complex surface X. Since D has normal crossings, we can assume that the divisor D has local equation $z_1 = 0$ away from an intersection point, and local equation $z_1 z_2 = 0$ near an intersection point, where z_1, z_2 are local holomorphic coordinates on X. The locally free sheaf $\Omega_X^1(\log D)$ is generated over the structure sheaf $\mathcal{O}_X$ of X by the holomorphic forms on X and the logarithmic differentials dz/z, where $z = 0$ is the local equation for a component of D. For $i = 1, 2$, the cohomology class

$$\bar{c}_i := (-1)^i c_i(\Omega_X^1(\log D))$$

is called the ith *logarithmic Chern class* of the pair (X, D). It is computed by analogy with the case $D = \phi$, where we recover the Chern classes of X. We have the corresponding *logarithmic Chern numbers* $\bar{c}_1^2$ and $\bar{c}_2$ that are given by

$$\bar{c}_1^2 = (K_X + D)^2$$

and

$$\bar{c}_2 = e(X) - e(D) = e(X \setminus D).$$

These formulas follow from the exact sequence

$$0 \to \Omega^1_X \to \Omega^1_X(\log D) \to \mathcal{O}_D \to 0.$$

The Chern forms $\gamma_1(X, D)$ and $\gamma_2(X, D)$ for (X, D) are computed using a smooth Hermitian metric of the dual sheaf $\Omega^1_X(\log D)^*$, in place of one for the tangent sheaf of X, in the formulas for the Chern forms of Chapter 4. We have

$$\bar{c}_1^2 = \int_X \gamma_1(X, D)^2, \qquad \bar{c}_2 = \int_X \gamma_2(X, D).$$

For more details about logarithmic forms, see [49], pp. 449–454, [68], [69], [70], Chapter 11. Recall that in Chapter 4, §4.3, and Chapter 5, §5.4, we encountered a pair (Y, C), where Y is a smooth compact surface and C is a smooth elliptic curve on Y with negative self-intersection. By the adjunction formula, we have

$$\bar{c}_1 = (K_Y + C)^2 = K_Y^2 - C \cdot C = c_1^2(Y) - C \cdot C$$

and

$$\bar{c}_2 = e(Y) - e(C) = c_2(Y) = e(Y \setminus C).$$

In Chapter 5, §5.4, we denoted in this case $\bar{c}_1$ by $c_1(Y, C)$ and $\bar{c}_2$ by $c_2(Y, C)$.

6.1 EXISTENCE OF FINITE COVERS BY BALL QUOTIENTS OF WEIGHTED CONFIGURATIONS: THE GENERAL CASE

A normal complex surface X has only isolated (normal) singularities, and, if X is projective, any resolution of its singularities is a smooth complex projective variety. A $\mathbb{Q}$-divisor D is a formal sum $\sum_i q_i D_i$ of finitely many irreducible curves D_i with coefficients q_i rational numbers. A *logarithmic pair*, or *log-pair*, is a pair (X, D) consisting of a normal variety X and a $\mathbb{Q}$-divisor D with $0 \leq q_i \leq 1$ for all i. The *log-canonical divisor* of a log-pair (X, D) is defined to be $K_X + D$, where K_X is the canonical divisor of X. In Chapter 3, we only defined the canonical divisor on a smooth compact algebraic surface. On a normal surface, we can first define the canonical divisor on its smooth locus, and then observe that this defines a unique divisor class on X. In the applications to line arrangements later in this chapter, we express all the divisors occurring on normal surfaces in terms of divisors on smooth compact algebraic surfaces. Therefore, for the moment, we do not give more background on divisors on normal surfaces. For details, see [10], Chapters 3 and 4.

In [83], [84] a log-pair (X, D), for X a compact complex normal surface and D a $\mathbb{Q}$-divisor of the form $D = \sum_i (1 - \frac{1}{b_i})D_i$, where the D_i are irreducible curves on X and the $b_i \geq 2$ are integers or ∞, is called a *normal surface pair*. We no longer necessarily assume that the D_i meet transversally. If X is smooth and the D_i meet transversally, we call (X, D) a *regular normal surface pair*. In practice, we will always express our $\mathbb{Q}$-divisors on normal surfaces as $\mathbb{Q}$-divisors on related smooth surfaces whose support has irreducible components with only transverse intersections. The intersection numbers are then defined as in Chapter 3, and extended by bilinearity from $\mathbb{Z}$-divisors to $\mathbb{Q}$-divisors.

Definition 6.1 *Let (X, D) be a regular normal surface pair. Let Z be an irreducible curve on X meeting the support of $K_X + D$ transversally. If $Z \cdot Z < 0$ and $(K_X + D) \cdot Z < 0$, then Z is called a log-exceptional curve of the first kind for (X, D). If $Z \cdot Z < 0$ and $(K_X + D) \cdot Z = 0$, then Z is called a log-exceptional curve of the second kind for (X, D).*

As we remark in the next paragraph, this definition can be adapted to make sense without the regularity assumption. Any curve with negative self-intersection on a normal surface contracts to a point on a normal surface ([124], p. 452). By successive contractions of all log-exceptional curves of the first kind on (X, D), we arrive at a *log-minimal surface pair* (X', D'), that is, it contains no log-exceptional curves of the first kind. We call (X', D') a *log-minimal model* of (X, D). If we further contract the images of the log-exceptional curves of the second kind in the log-minimal model (X', D'), we obtain a *log-canonical surface pair* (X'', D''), that is, it contains neither log-exceptional curves of the first kind nor log-exceptional curves of the second kind. We call (X'', D'') a *log-canonical model* of (X, D). When $D = \phi$, we recover the notions of minimal model and canonical model of Chapter 4.

For the applications to line arrangements in $\mathbb{P}_2$ that we discuss in this chapter, the log-exceptional curves will be pairwise disjoint. Let (X, D) be a regular normal surface pair, and let $f : X \to X^1$ be the contraction of a disjoint union C of log-exceptional curves. Let

$$D^1 = f_*(D) := \sum_i \left(1 - \frac{1}{b_i}\right) f(D_i),$$

where $f(D_i)$ is omitted if D_i appears in the disjoint union C. Then, we define the pullback $f^*(K_{X^1} + D^1)$ of $K_{X^1} + D^1$ to be the $\mathbb{Q}$-divisor on X satisfying

$$f^* \left(K_{X^1} + D^1\right) \cdot C = 0.$$

It follows that $f^*(K_{X^1} + D^1)$ is defined by the equation

$$K_X + D = f^* \left(K_{X^1} + D^1 \right) + \frac{(K_X + D) \cdot C}{C^2} \, C. \tag{6.1}$$

This gives an explicit expression for the lift of the log-canonical divisor of (X^1, D^1) in terms of the log-canonical divisor of (X, D), a formula we use frequently in what follows, although we drop the reference to the map f. In other words, we identify the log-canonical divisor $K_{X^1} + D^1$ on (X^1, D^1), whose support does not necessarily have normal crossings, with the $\mathbb{Q}$-divisor

$$K_X + D - \frac{(K_X + D) \cdot C}{C^2} \, C$$

on X and use this divisor, which has normal crossings, in our formulas involving intersection numbers. This is a special case of the intersection theory for normal surfaces developed by Mumford [111]; see also [123], [124]. Let Y be a normal compact complex surface and $\pi : \overline{Y} \to Y$ a resolution of singularities with exceptional set $E = \cup E_i$, where the E_i are the irreducible components of E. Let $\overline{D}$ be the strict transform of a divisor D on Y. Then the inverse image $\pi^* D$ of D is given by the $\mathbb{Q}$-divisor

$$\pi^* D = \overline{D} + \sum_i a_i E_i,$$

where the a_i are rational numbers determined by $(\pi^* D) \cdot E_j = 0$ for all the components E_j of E, namely

$$\sum_i a_i (E_i \cdot E_j) = -\overline{D} \cdot E_j.$$

These rational numbers exist as the intersection matrix $(E_i \cdot E_j)$ is negative definite. If D and D' are two divisors on X, then $D \cdot D'$ is defined to be the rational number $(\pi^* D) \cdot (\pi^* D')$. For more details, see [84], §3.2.1.

The pair (X'', D'') will in general have isolated singularities, and this will be relevant to our discussion of the proof of Theorem 6.4 below. Let (X, D) be a normal surface pair. A point p is defined to be a singularity of (X, D) if either p is a singular point of X or p is a smooth point of X but the curve $\mathrm{Supp}(D)$, which denotes the support of D, has a singularity at p. A singular point is often called *quasi-regular* if it is a smooth point of X. By a *regular point*, we shall mean a point that is a singularity neither of X nor of $\mathrm{Supp}(D)$. Let (V, D, p) be a germ of a normal surface pair, that is, (V, p) is a germ of a normal surface and $D = \sum_i (1 - \frac{1}{b_i}) D_i$, where the b_i are integers at least 2 or infinity. Moreover, each point D_i passes through the point p. A *resolution*

$(\widetilde{V}, \widetilde{D}, \widetilde{E})$ of (V, D, p) consists of

- a resolution $\pi : (\widetilde{V}, \widetilde{E}) \to (V, p)$ such that the proper transform $\cup_i \widetilde{D}_i$ of $\cup_i D_i$ is nonsingular,
- $\widetilde{D} = \sum_i (1 - \frac{1}{b_i}) \widetilde{D}_i$,
- $\widetilde{E} = \pi^{-1}(p)$.

A resolution of (V, D, p) is *good* if the union of $\widetilde{E}$, together with the proper transforms of the support of D, has only simple normal crossings. If we successively blow down the -1 curves that preserve goodness, we arrive at a *minimal good resolution*. The blow-ups of the multiple intersection points of our line arrangements give rise to minimal good resolutions. Associated to (V, D) we have the log-canonical divisor $K_V + D$. If

$$\pi : (\widetilde{V}, \widetilde{E}) \to (V, p)$$

is a resolution, then

$$\pi^* (K_V + D) = K_{\widetilde{V}} + \widetilde{D} + \Delta,$$

where the support of Δ is a subset of $\widetilde{E} = \cup_\alpha E_\alpha$, where the E_α are irreducible curves with normal crossings. Moreover, $\Delta = \sum_\alpha a_\alpha E_\alpha$ for rational numbers a_α determined by the equations ([111], [123], [124])

$$\left(K_{\widetilde{V}} + \widetilde{D} + \Delta \right) \cdot E_\alpha = 0.$$

Compare this to our discussion above of lifting the log-canonical divisor to a regular normal surface pair.

Definition 6.2 *A germ of a normal surface pair (V, D, p) is a log-canonical singularity if there exists a good resolution such that $a_\alpha \leq 1$ for all α. A log-canonical singularity is called log-terminal if $a_\alpha < 1$ for all α and $b_i < \infty$ for all i. A log-canonical singularity is LCS if it is log-canonical and not log-terminal ([84], p.337).*

Although we shall only need the results of [83], [84] in a special case, we now quote the more general result of those references since we have all the vocabulary for its statement. The proof of this general result follows lines similar to that of our Theorem 6.4, the main difference for the general normal surface pair being the wider range of possibilities for the singularities on the log-canonical model. Let (X, D) be a normal surface pair with at worst log-canonical singularities, where $D = \sum_i \left(1 - \frac{1}{b_i} \right) D_i$, the b_i either are positive integers greater than or equal to 2 or are ∞, and the D_i are the irreducible components of the support of D.

Let (X'', D'') be its log-canonical model, which also has at worst log-canonical singularities. Let D_i'' denote the image of D_i in X''. The *log-Kodaira dimension* $\overline{\kappa}(X, D)$ is defined in an analogous way to the usual Kodaira dimension of Chapter 4 by replacing the canonical divisor K_X by $K_X + D$ (see [84], §3.0, and [123], [124]). Of course X and X'' may no longer be smooth. As we said already, they have a suitable divisor and intersection theory and, in any case, we always use the strategy of working with the lifts of our divisors on normal surfaces to divisors with transverse intersection points on a smooth surface. The assumption in [83], [84] is that $\overline{\kappa}(X, D) = 2$. Later on, we just use the consequence needed, namely that the log-canonical divisor $K_{X''} + D''$ is ample, in that a positive integer multiple of it is very ample, and realizes X'' inside a projective space.

The following result is due to R. Kobayashi, S. Nakamura, and F. Sakai, and is [84], Theorem 1. We explain the word "orbifold" occuring there after the statement of Theorems 6.1 and 6.2.

Theorem 6.1 *Let (X, D) be a normal surface pair with only log-canonical singularities and $\overline{\kappa}(X, D) = 2$. Let (X'', D'') be its log-canonical model. Then,*

(i) *if $X_0'' = X'' \setminus \cup_{b_i=\infty} D_i'' \setminus \mathrm{LCS}(X'', D'')$ and $D_0'' = D'' \cap X_0''$, then (X_0'', D_0'') is an orbifold with branch loci the support of D_0'' and branch index b_i along $D_{0,i}'' = D_i'' \cap X_0''$; in particular, any singular point of X'' not contained in $\cup_{b_i=\infty} D_i'' \cup \mathrm{LCS}(X'', D'')$ is an isolated quotient singularity;*

(ii) *there exists a unique complete Kähler-Einstein orbifold metric with negative scalar curvature on the orbifold (X_0'', D_0'') whose Kähler form ω gives a real closed $(1, 1)$-current on any resolution $\mu : Y'' \to X''$ with cohomology class satisfing $[\mu^*(\omega)] = 2\pi c_1(\mu^*(K_{X''} + D''))$.*

For the definition of a real closed $(1, 1)$-current, see [49], Chapter 3, §1. The form ω will in general be singular on X_0'' but it lifts to a nonsingular form on the local orbifold charts of (X_0'', D_0''). The corresponding unique complete Kähler-Einstein orbifold metric on the log-canonical model (X'', D'') is the canonical Kähler-Einstein metric of (X, D). On integrating the Chern forms for the canonical Kähler-Einstein metric of (X, D) in Theorem 6.1, we deduce the following "Miyaoka-Yau inequality" for the *(orbifold) logarithmic Chern numbers* $\overline{c}_1(X'', D'')^2$ and $\overline{c}_2(X'', D'')$, quoted here from [84], Theorem 2, and again due to R. Kobayashi, S. Nakamura, and F. Sakai. These are related to, but should not be confused with, the $\overline{c}_1^2$ and $\overline{c}_2$ introduced earlier.

Theorem 6.2 *Let (X, D) and (X'', D'') be as in Theorem 6.1. Then we have*

$$\overline{c}_1(X'', D'')^2 \le 3\overline{c}_2(X'', D''),$$

where

$$\bar{c}_1(X'', D'')^2 = \left(K_{X'} + D''\right)^2$$

and

$$\bar{c}_2(X'', D'') = e(X_0'') + \sum_i \left(\frac{1}{b_i} - 1\right) \left(e(D_{0,i}'') - d_i\right) + \sum_P \left(\frac{1}{|\Gamma(P)|} - 1\right),$$

where P runs over the log-terminal singular points of (X'', D''). Here $e(X_0'')$ is the Euler number of X_0''. The integer d_i is the number of singularities of (X'', D'') that lie on $D_{0,i}''$, and $|\Gamma(P)|$ is the order of the local uniformizing group (which is a finite subgroup of $U(2)$) of a local orbifold chart centered at the log-terminal singular point P of (X'', D''). We have the equality

$$\bar{c}_1(X'', D'')^2 = 3\bar{c}_2(X'', D'')$$

if and only if the orbifold (X_0'', D_0'') is biholomorphic to the ball quotient $\Gamma \backslash B_2$ with Γ a discrete subgroup of $\mathrm{PU}(1, 2)$ and $\cup_{b_i \neq \infty} D_{0,i}''$ is the branch loci of the natural map $B_2 \rightarrow X_0''$, with branch index b_i along $D_{0,i}''$.

Unlike in the situation of Chapter 4, the lattice Γ in $\mathrm{PU}(1, 2)$ appearing in Theorem 6.2 in general has fixed points on B_2. Indeed, the theorem shows the existence of a lattice Γ in $\mathrm{PU}(1, 2)$ such that the natural Γ-invariant map $B_2 \rightarrow \Gamma \backslash B_2$ has image isomorphic to X_0'' and is ramified of order $b_i > 0$ along $D_i'' \cap X_0''$. When we look for a Kähler-Einstein metric on X_0'' that will ultimately agree with the metric induced on $\Gamma \backslash B_2$ by the ball metric, we must expect this metric to have singularities along D_0''. This leads to the notion of "orbifold" metric or "b-metric", which we explain in the next section.

6.2 REMARKS ON ORBIFOLDS AND b-SPACES

The notion of *orbifold* is useful when we consider uniformizations, local or global, with the covering group acting properly discontinuously with fixed points, as opposed to without fixed points when the notion of manifold suffices. An orbifold consists of a (possibly singular) complex manifold X (in our case an algebraic variety) of dimension n, with only finite quotient singularities. This means that the singularities of X are all locally isomorphic to quotient singularities $\mathbb{C}^n / G$ for G a finite subgroup of $\mathrm{GL}(n, \mathbb{C})$ (acting effectively: for any two distinct g, h in G there exists a $u \in \mathbb{C}^n$ such that $g(u) \neq h(u)$). This means that X can be covered by *orbifold charts*. The charts centered at a nonsingular point are just the usual manifold charts. The charts centered at a singular point x of X are of the form (U, G, φ), where U is a neighborhood of the origin 0 in $\mathbb{C}^n$, where G is a finite subgroup of $\mathrm{GL}(n, \mathbb{C})$, unique up to conjugation, such that the open neighborhoods of x in X and the origin 0 in U/G can be identified. The gluing condition on charts is the following. If $V' \subset V$ are open sets in X with charts $U'/G' \simeq V'$ and $U/G \simeq V$, then there should exist a (nonunique) monomorphism $G' \hookrightarrow G$ and an injection $U' \hookrightarrow U$ commuting with the given G'-action on U' and its action through $G' \hookrightarrow G$ on U. It follows from the gluing condition that the order of a point $x \in X$, that is, the cardinality of the stabilizer of any lift of x in any orbifold chart, is well defined. The orbifold locus, or singular set of the orbifold, is the set of points with a nontrivial stabilizer group. The orbifold locus need not be isolated. Clearly, if Y is a smooth complex manifold and G is a finite group acting holomorphically on Y, then Y/G is a complex orbifold. For each $y \in Y$, the point Gy of Y/G is singular if and only if the stabilizer G_y of y in G is nontrivial. For more details, see [21], [46], [72], and [120].

We do not need the most general definition of an orbifold, but rather one that applies to the complex normal surface pairs of this chapter. Also, we are not absolutely rigorous in that we do not worry about the gluing conditions, for example. Following the treatment in [138], we use the related notion of *b-space*, which we now explain, since it appears in our key reference by Kobayashi [84], where, as in §6.1, the orbifold structure is given by a $\mathbb{Q}$-divisor of the form $\sum_i (1 - \frac{1}{b_i}) D_i$. In our case, the b_i are determined by the weights assigned to the lines of our arrangements in $\mathbb{P}_2$.

A *transformation group* is a pair $(\Gamma; \widehat{U})$ where $\widehat{U}$ is an open subset of a connected complex manifold and Γ is a group of holomorphic automorphisms of $\widehat{U}$ acting properly discontinuously (here we do not assume that Γ is finite). In particular, the set $\widehat{U}$ is invariant under Γ and, for all $z \in \widehat{U}$, the *isotropy group* at z is defined by

$$\Gamma_z = \{\gamma \in \Gamma : \gamma(z) = z\},$$

and is finite. Let U be the orbit space $\Gamma \backslash \widehat{U}$ and $\pi : \widehat{U} \to U$ the natural projection. Then U has a well-defined map $b_\pi : U \to \mathbb{N}$ defined by

$$b_\pi : x \to \mathrm{Cardinality}\left(\Gamma_{\pi^{-1}(x)}\right),$$

for any choice of $\pi^{-1}(x)$. The singularities of U are all quotient singularities. This is the notion of *b-space* in the sense of Kato [75], namely a normal complex space X together with a map b from X to the nonnegative integers. We consider at times b-spaces with b-maps having an image in $\mathbb{N} \cup \{\infty\}$, but we always modify this space so as to consider a b-space whose b-map has finite values. Indeed, the most important example for us is when $\widehat{U}$ is the symmetric space given by the complex 2-ball B with a certain radius. Then Γ_z will be finite for all points in the interior of B, but we also consider infinite isotropy groups of points on the boundary of such a ball. It turns out that the nature of the finite-valued b-map on $\Gamma \backslash B$ determines whether Γ has fixed points on the boundary of B. If this is the case, then we may compactify our b-space by adding the orbits of such points. A b-space is said to be locally finitely uniformizable if every point $x \in X$ has a neighborhood U that is the orbit space of a transformation group $(\widehat{U}, \Gamma)$, where Γ is a finite group, and $b_\pi = b_{|U}$, where $\pi : \widehat{U} \to U$ is the natural projection. Of course, we are allowing the case where $\widehat{U} = U$ and Γ consists only of the identity map. A locally finite uniformizable b-space $(X; b)$ is an orbifold. The space X is said to be the base space of $(X; b)$, and $(X; b)$ is said to be an orbifold or b-space over X. The locus where the b-map is nontrivial is the orbifold or singular locus, which is also called the ramification or branching locus. A global (finite) uniformization of $(X; b)$ is a transformation group $(\Gamma; M)$, with Γ finite and with orbit space X, such that if $\pi : M \to X$ is the natural projection, then $b = b_\pi$. The existence of global uniformizations *that are free ball quotients* is the central question of this chapter. It turns out that the groups appearing in the local uniformizing charts are determined up to conjugation by the data given by the b-map, which is derived from the weights assigned to the blow-up of the weighted line arrangement in $\mathbb{P}_2$. Roughly speaking, for b-spaces coming from the log-canonical models (X'', D'') of the weighted blown-up line arrangements, the finite groups G will be cyclic (away from intersection points of D''), a product of cyclic groups (at transverse intersections points of D''), or (at multiple intersection points of D'') a member of an explicit list we shall give later in this chapter, after the proof of Theorem 6.4.

Any local object (e.g., a tensor) on a b-space is defined to be a Γ-invariant object on a local chart $(\widehat{U}, \Gamma)$ given by a transformation group "upstairs" (rather than an object "downstairs" on the underlying space). Therefore a Kähler b-metric, or orbifold metric, is a Γ- invariant Kähler metric on $\widehat{U}$ for each orbifold chart $\widehat{U} \to \widehat{U}/\Gamma$ that glues: its pullback under an injection $\widehat{U}' \hookrightarrow \widehat{U}$ of charts "upstairs" is the corresponding metric on $\widehat{U}'$. Such a metric descends to give a Kähler metric on the underlying space "downstairs," but with possible singularities along the orbifold locus. Similarly, as continuity and differentiability are local properties, one can talk of b-C^k functions on a b-space, as in the work of Kato [75].

6.3 $K_{X''} + D''$ FOR WEIGHTED LINE ARRANGEMENTS

We return now to a weighted line arrangement of $k \geq 6$ lines $L_1, \ldots, L_k$ in the complex projective plane $\mathbb{P}_2$ and derive a formula (namely (6.7) below) for the divisor that plays the role of the divisor $K_{X''} + D''$ of Theorem 6.2 for this situation. Along the way, we introduce some other $\mathbb{Q}$-divisors on X that lift, on the proposed ball quotient finite covers, to the canonical divisors encountered in the discussion of Chapter 5.

Recall that we always express our $\mathbb{Q}$-divisors on normal surfaces in terms of $\mathbb{Q}$-divisors on smooth surfaces whose support has irreducible components with only transverse intersections. The intersection numbers are then defined as in Chapter 3, extended by bilinearity from $\mathbb{Z}$-divisors to $\mathbb{Q}$-divisors.

As before, the weights are nonzero integers, or infinity, denoted by $n_1, \ldots, n_k$ for the lines and m_j for the singular points p_j of the arrangement. The projective plane is blown up at the singular points p_j. On the resulting surface X, we have smooth rational curves given by the proper transforms D_i of the L_i, $i = 1, \ldots, k$, and curves E_j, which are the blow-ups of the p_j. The D_i have weights n_i and the E_j have weights m_j.

We always assume that the D_i and E_j whose weights are negative or infinity are mutually disjoint on X.

To the weights n_i, m_j we associate n'_i, m'_j, where n'_i equals n_i if $n_i > 0$, equals $-n_i$ if $n_i < 0$, and equals an arbitrary positive integer if $n_i = \infty$. The m'_j are defined in the same way. Let

$$D = \sum_{i=1}^{k} \left(1 - \frac{1}{n'_i}\right) D_i + \sum_j \left(1 - \frac{1}{m'_j}\right) E_j. \tag{6.2}$$

The regular normal surface pair (X, D) has log-canonical divisor $K_X + D$ where

$$K_X = \pi^* K(\mathbb{P}_2) + \sum_j E_j,$$

for $\pi : X \to \mathbb{P}_2$ the natural map. (From the discussion in Chapter 3 it is clear that, if there exists a good cover $Y \to X$ with ramification indices n'_i along D_i and m'_j along E_j, then $K_X + D$ lifts to the canonical divisor of Y.)

We recall the diophantine conditions introduced in Chapter 5, §5.2 (that is, the vanishing of P_j in (5.14) and Q_i in (5.15)):

$$\frac{2}{m_j} = r_j - 2 - \sum_{D_i \cap E_j \neq \phi} \frac{1}{n_i},$$

$$\frac{2(\sigma_i - 1)}{n_i} = \tau_i - 2 - \sum_{E_j \cap D_i \neq \phi} \frac{1}{m_j} - \sum_{D_\ell \cap D_i \neq \phi} \frac{1}{n_\ell}. \tag{6.3}$$

We assume that these diophantine conditions are satisfied for divisors whose weights appearing on the left-hand sides are negative or infinite. Recall that r_j is the number of lines passing through the point p_j (whose blow-up is E_j) and τ_i is the total number of intersection points on L_i, whereas σ_i is the total number of singular intersection points on L_i.

For E_s with negative weight $m_s = -m'_s$, we have

$$(K_X + D) \cdot E_s = K_X \cdot E_s + \sum_{i \sim s} \left(1 - \frac{1}{n'_i}\right) D_i \cdot E_s + \left(1 - \frac{1}{m'_s}\right) E_s \cdot E_s$$

$$= -2 - E_s \cdot E_s + \sum_{i \sim s} \left(1 - \frac{1}{n_i}\right) + E_s \cdot E_s - \frac{1}{m'_s} E_s \cdot E_s$$

$$= r_s - 2 - \sum_{i \sim s} \frac{1}{n_i} - \frac{1}{m'_s} E_s \cdot E_s = \frac{2}{m_s} + \frac{E_s \cdot E_s}{m_s} = \frac{1}{m_s}.$$

In this formula, the index $i \sim s$ ranges over those i with $D_i \cap E_s \neq \phi$. Notice that, in the second line, we have applied the adjunction formula to deduce $K_X \cdot E_s = -2 - E_s \cdot E_s$. In the last line, we have used the first diophantine condition from (6.3). As $E_s \cdot E_s = -1$, and $m_s < 0$, we see from this last equation that E_s is a log-exceptional curve of the first kind for (X, D). Moreover, by (6.1), the contraction $f : X \to X^1$ of E_s gives a normal surface pair (X^1, D^1) with $D^1 = f_* D$ and

$$K_{X^1} + D^1 = K_X + D + \frac{1}{m_s} E_s = K_X + D - \frac{1}{m'_s} E_s. \qquad (6.4)$$

Similarly, for D_t with negative weight $n_t = -n'_t$, we have

$$(K_X + D) \cdot D_t = K_X \cdot D_t + \sum_{\ell \sim t} \left(1 - \frac{1}{n'_\ell}\right) D_\ell \cdot D_t + \left(1 - \frac{1}{n'_t}\right) D_t \cdot D_t$$

$$+ \sum_{j \sim t} \left(1 - \frac{1}{m'_j}\right) E_j \cdot D_t$$

$$= -2 - D_t \cdot D_t + D_t \cdot D_t + \frac{1}{n_t}(1 - \sigma_t) + \tau_t - \sum_{\ell \sim t} \frac{1}{n_\ell} - \sum_{j \sim t} \frac{1}{m_j}$$

$$= \tau_t - 2 - \sum_{\ell \sim t} \frac{1}{n_\ell} - \sum_{j \sim t} \frac{1}{m_j} + \frac{1 - \sigma_t}{n_t} = \frac{2(\sigma_t - 1)}{n_t} + \frac{1 - \sigma_t}{n_t} = \frac{\sigma_t - 1}{n_t}.$$

In this formula, the index $\ell \sim t$ runs over those ℓ with $D_\ell \cap D_t \neq \phi$, $\ell \neq t$, and the index $j \sim t$ runs over those j with $E_j \cap D_t \neq \phi$. Again, in the second line, we have applied the adjunction formula to deduce that $K_X \cdot D_t = -2 - D_t \cdot D_t$, and in the last line we have this time used the second diophantine condition from (6.3). Recall that $D_t \cdot D_t = 1 - \sigma_t$ and that $\sigma_t \geq 2$. Therefore D_t is a log-exceptional curve of the first kind for (X, D). Moreover, by (6.1), the contraction $g : X \to X^1$ of D_t gives a normal surface pair (X^1, D^1) with $D^1 = g_* D$ and

$$K_{X^1} + D^1 = K_X + D + \frac{1}{n_t} D_t = K_X + D - \frac{1}{n'_t} D_t. \qquad (6.5)$$

Let $f : X \to X'$ be the contraction of all the curves E_s with $m_s < 0$ and D_t with $n_t < 0$ and let $\mathcal{D}' = f_* \mathcal{D}$. Then, by (6.4) and (6.5) we have

$$K_{X'} + \mathcal{D}' = K_X + \mathcal{D} - \sum_{m_s < 0} \frac{E_s}{m'_s} - \sum_{n_t < 0} \frac{D_t}{n'_t}. \qquad (6.6)$$

Notice that all the E_s and all D_t with $\sigma_t = 2$ have self-intersection -1, so their contractions give rise to smooth points on X'. However, the D_t with $\sigma_t \geq 3$ give rise to singular points on X'. We will discuss this in more detail later. (If the good covering $Y \to X$ referred to previously exists, then the $\mathbb{Q}$-divisors $\sum_{m_s < 0} E_s / m'_s$ and $\sum_{n_t < 0} D_t / n'_t$, when lifted to Y, are a union of exceptional curves. Let Y' be the smooth surface obtained from Y by collapsing all these exceptional curves. Then $K_{X'} + \mathcal{D}'$ would lift to the canonical divisor of Y'.)

We now turn to the divisors E_s and D_t with infinite weight. We again use the adjunction formula and the diophantine conditions (6.3), and the same notation for $i \sim s, i \sim t, j \sim t$. For $m_s = \infty$, we have

$$(K_X + \mathcal{D}) \cdot E_s = r_s - 2 - \sum_{i \sim s} \frac{1}{n_i} - \frac{1}{m'_s} \qquad E_s \cdot E_s = \frac{1}{m'_s},$$

and for $n_t = \infty$, we have

$$(K_X + \mathcal{D}) \cdot D_t = \tau_2 - 2 - \sum_{\ell \sim t} \frac{1}{n_\ell} - \sum_{j \sim t} \frac{1}{m_j} - \frac{(1 - \sigma_t)}{n'_t} = \frac{(\sigma_t - 1)}{n'_t}.$$

As we have assumed no two divisors of infinite or negative weight intersect, it follows that we also have

$$(K_{X'} + \mathcal{D}') \cdot E_s = \frac{1}{m'_s}, \qquad (K_{X'} + \mathcal{D}') \cdot D_t = \frac{(\sigma_t - 1)}{n'_t}.$$

Let

$$C' = \sum_{m_s = \infty} \frac{E_s}{m'_s} + \sum_{n_t = \infty} \frac{D_t}{n'_t},$$

so that $D = \mathcal{D} + C'$ is given by

$$D = \sum_{n_i \neq \infty} \left(1 - \frac{1}{n'_i}\right) D_i + \sum_{m_j \neq \infty} \left(1 - \frac{1}{m'_j}\right) E_j + \sum_{n_i = \infty} D_i + \sum_{m_j = \infty} E_j.$$

Then (X, D) is a normal surface pair, where we have chosen the notation D for the $\mathbb{Q}$-divisor above to match that of [83], [84]. As the divisors E_s and D_t of negative and infinite weight are pairwise disjoint, those of negative weight are log-exceptional curves of the first kind for (X, D), as they are log-exceptional curves of the first kind for $(X, \mathcal{D})$ and $C' \cdot E_s = C' \cdot D_t = 0$. For the E_s and D_t of infinite weight we have,

$$(K_X + D) \cdot E_s = (K_X + \mathcal{D} + C') \cdot E_s = \frac{1}{m'_s} + \frac{E_s \cdot E_s}{m'_s} = \frac{1}{m'_s} - \frac{1}{m'_s} = 0,$$

and

$$(K_X + D) \cdot D_t = (K_X + \mathcal{D} + C') \cdot D_t$$

$$= \frac{(\sigma_t - 1)}{n'_t} + \frac{D_t \cdot D_t}{n'_t} = \frac{(\sigma_t - 1)}{n'_t} + \frac{(1 - \sigma_t)}{n'_t} = 0.$$

Therefore, the E_s and D_t of infinite weight are log-exceptional curves of the second kind for (X, D). Similarly, if $D' = \mathcal{D}' + C'$, then (X', D') is a normal surface pair and we have $(K_{X'} + D') \cdot E_s = (K_{X'} + D') \cdot D_t = 0$ for the E_s, D_t with infinite weight. Recall that for the E_s and D_t of negative weight we also have $(K_{X'} + D') \cdot E_s = (K_{X'} + D') \cdot D_t = 0$. If we contract the images of the E_s and D_t of infinite weight on X', we get a new normal surface X'' and, by (6.1), the divisor $K_{X'} + D'$ descends as is to X'', where it has the form $K_{X''} + D''$ for a $\mathbb{Q}$-divisor D'' on X''. We may, and will where convenient, view this as a $\mathbb{Q}$-divisor on X given explicitly by the following formula, which depends only on the given weighted line arrangement:

$$K_{X''} + D'' = \pi^* K(\mathbb{P}_2) + \sum_{i=1}^{k} \left(1 - \frac{1}{n_i}\right) D_i + \sum_j \left(2 - \frac{1}{m_j}\right) E_j$$

$$+ 3 \sum_{n_i < 0} \frac{D_i}{n_i} + 3 \sum_{m_j < 0} \frac{E_j}{m_j}. \quad (6.7)$$

Again, we stress that the $\mathbb{Q}$-divisor $K_{X''} + D''$ on X has zero intersection with the E_s and D_t of negative and infinite weight. (If the good covering $Y \to X$ referred to previously exists, the divisor C' lifts to Y. If C denotes the image of C' in Y', then $K_{X''} + D''$ lifts to a divisor $K(Y', C)$ on $Y' \setminus C$ with self-intersection given by $c_1^2(Y', C)$, as in Chapter 5, §5.4.)

6.4 EXISTENCE QUESTION

We approach the problem of the *existence* of ball quotient finite coverings $Y' \setminus C$ of a line arrangement, where Y' and C are as described in §6.3 and Chapter 5, by first showing the *existence* of a discrete subgroup of $PU(1, 2)$ that ensures that the ball is an appropriately ramified infinite cover of X''. More precisely, we examine the following problem, which is a special case of that considered in Theorems 6.1 and 6.2 above, applied to the normal projective variety X'' derived from a weighted line arrangement:

Question: *When is X'' a (possibly compactified) ball quotient $X'' = \overline{\Gamma \backslash B_2}$ for a discrete subgroup Γ of the automorphisms of the ball with natural map $B_2 \to X''$ having branching of order n_i and m_j respectively along images in X'' of the D_i and E_j with $n_i > 0$ and $m_j > 0$?*

As we shall discuss in more detail later, any normal subgroup Γ' of finite index N in Γ acting on B_2 without fixed points gives rise to a free ball quotient $Y' \backslash C = \Gamma' \backslash B_2$, as described in Chapter 5, §5.4, which is a finite cover of $X_0'' := \Gamma \backslash B_2$ of degree N, ramified of order $n_i > 0$ along the D_i and $m_j > 0$ along the E_j. Such normal subgroups exist by work of Borel [18] and Selberg [129]. Therefore, the existence of Γ gives the existence of suitable covers $Y' \backslash C$. Here C is a union of smooth elliptic curves on Y' resolving the set of cusp singularities $X'' \backslash X_0''$, which is nonempty if Γ has fixed points on the boundary of B_2, each cusp coming from such a point (whose isotropy group is always infinite). We shall see that the points in X'' corresponding to the E_s with $m_s < 0$ and to the D_t with $n_t < 0$ are quotient singularities with respect to finite isotropy subgroups of Γ. The points corresponding to the E_s with $m_s = \infty$ and D_t with $n_t = \infty$ give rise to the cusps.

6.5 AMPLENESS OF $K_{X''} + D''$

One of the conditions needed for an affirmative answer to the question of §6.4 is the ampleness of the divisor $K_{X''} + D''$, which, in the general case, is a consequence of the assumption $\overline{\kappa}(X, D)$ of Theorem 6.1. We need this ampleness criterion for $K_{X''} + D''$ as part of the construction of the Kähler-Einstein b-metric on (X_0'', D_0''). We will use a positive integer multiple of this ample divisor to map X to a surface, isomorphic to X'', in a projective space and then pull back to X the Fubini-Study metric on this projective space.

Recall that we can view $K_{X''} + D''$ as a $\mathbb{Q}$-divisor on X that has zero intersection with the divisors D_s and E_t of negative and infinite weight. By the self-intersection number $(K_{X''} + D'')^2$ we mean the self-intersection number of the divisor on X given by the right-hand side of (6.7). This is consistent with our remarks about intersection numbers of divisors on normal surfaces in §6.1, where the normal surface here is X'' and X is a resolution of the singularities of X''. If S is a divisor on X, then by $(K_{X''} + D'') \cdot S$ we mean the intersection number of the divisor on X given by the right-hand side of (6.7) and S.

A $\mathbb{Q}$-divisor on a normal surface is called numerically ample if its self-intersection number is positive and if its intersection with all irreducible curves on the surface is also positive. This terminology has its origin in the Nakai-Moishezon ampleness criterion [107], [113] (see [51], p. 365, for an exposition for smooth projective surfaces). By Corollary (9) in [83], if $K_{X''} + D''$ is effective (the coefficients of the irreducible components of its support are all nonnegative) and numerically ample, then it is ample and $\overline{\kappa}(X, D) = 2$; see also [84], Fact 4, p. 352. Here, as before, the notation $\overline{\kappa}(X, D)$ refers to the *log-Kodaira dimension*, which is defined in the same way as the Kodaira dimension in Chapter 4, §4.1, with K_X now replaced by $K_{X''} + D''$. These conditions ensure the appropriate "general type" property of (X, D), which is necessary for the corresponding Miyaoka-Yau inequality to hold.

Suppose that $X'' = \overline{\Gamma \backslash B_2}$ as in Theorem 6.2 of §6.1. Suppose $K_{X''} + D''$ corresponds to a Kähler form, as in Theorem 6.1 of §6.1, coming from the unique, up to a nonzero scalar, invariant Kähler form on B_2. It follows that $K_{X''} + D''$ is effective. We have $(K_{X''} + D'')^2 > 0$ as this number is essentially the volume of X''. Moreover, for any curve S'' on X'', we have $(K_{X''} + D'') \cdot S'' > 0$. We shall see how, conversely, these conditions, which as we have seen imply the ampleness of $K_{X''} + D''$, together with equality in a Miyaoka-Yau inequality (given by Theorem 6.2 specialized to our case), are equivalent to the condition $\mathrm{Prop}(Y', C) = 0$ on the sought-after $Y' \backslash C$ of Chapter 5, §5.4, and guarantee that X'' is an appropriately ramified ball quotient.

As we have just found an explicit formula for $K_{X''} + D''$, it is convenient to now determine conditions on the weights that ensure these ampleness criteria and check that they are fulfilled by the weights on the line arrangements giving $\mathrm{Prop}(Y', C) = 0$ listed in Chapter 5. Firstly, we determine conditions on the weights that ensure that $K_{X''} + D''$ is linearly equivalent to an effective divisor on X. We can take $K(\mathbb{P}_2)$ to be $-\frac{3}{k}(L_1 + \ldots + L_k)$ and, as L_i lifts to $D_i + \sum_{D_i \cap E_j \neq \phi} E_j$ on X, we deduce from (6.7) that, up to linear equivalence,

$$
K_{X''} + D'' = \sum_{i=1}^{k} \left(1 - \frac{1}{n_i} - \frac{3}{k} \right) D_i + \sum \left(2 - \frac{1}{m_j} - \frac{3}{k} r_j \right) E_j
$$
$$
+ 3 \sum_{n_i < 0} \frac{D_i}{n_i} + 3 \sum_{m_j < 0} \frac{E_j}{m_j}. \quad (6.8)
$$

In (6.8), the coefficient of D_i is nonnegative either if $n_i \geq 2$, $n_i = \infty$ and $k \geq 6$, or, for negative n_i, if $n_i \leq -4$ and $k \geq 6$. The coefficient of E_j is nonnegative either if $m_j > 0$, including $m_j = \infty$ and $r_j \leq k/3$, or if $m_j \geq 2$ and $r_j \leq k/2$. For negative m_j, the coefficient of E_j is non-negative if $m_j \leq -4$ and $r_j \leq k/2$. Notice that, up to now, we only used the diophantine conditions (6.3) for the curves E_s or D_t with m_s or n_t negative or infinite.

Using the tables at the end of Chapter 5, we can check that $K_{X''} + D''$ is effective for all the line arrangements considered in that chapter with *all* the weights n_i and m_j satisfying the diophantine conditions. More precisely, for the first twenty-seven complete quadrilateral cases considered in Table 5.13, for the Hesse cases in Table 5.2, the icosahedral cases in Table 5.3, the G_{168} cases in Table 5.4, the A_6 cases in Table 5.5, and the extended Hesse arrangement cases in Table 5.11, one checks directly from (6.8) that the coefficients of the D_i and E_j in $K_{X''} + D''$ are nonnegative. Indeed, they are even all positive, except for two cases. Firstly, for the complete quadrilateral, the D_i and E_j of weights 2 and -4 have coefficients equal 0. Secondly, when all weights equal 5 in the icosahedral arrangement, the coefficients of all the E_j coming from the six blown-up quintuple points equal 0. In this case $K_{X''} + D''$ is still positive overall in that there are D_i or E_j with positive coefficients (the other D_i or E_j having coefficients equal to 0). We can also check directly that $K_{X''} + D''$ is effective for the additional cases in §5.6.1. We discuss Ceva(q) and $\overline{\text{Ceva}}(q)$ at the end of this section.

We now check that $(K_{X''} + D'')^2 > 0$ for the weighted line arrangements of Chapter 5. We use the representation of the linear equivalence class $K(\mathbb{P}_2) = -3H$, where $\pi^* H = H = D_i + \sum_{D_i \cap E_j \neq \phi} E_j$.

From (6.7) we have

$$
K_{X''} + D'' = \left(k - 3 - \sum_{i=1}^{k} \frac{1}{n_i} \right) H + \sum \left(2 - \frac{1}{m_j} - \sum_{D_i \cap E_j \neq \phi} \left(1 - \frac{1}{n_i} \right) \right) E_j
$$

$$
+ 3 \sum_{n_i < 0} \frac{D_i}{n_i} + 3 \sum_{m_j < 0} \frac{E_j}{m_j}. \tag{6.9}
$$

If we assume the diophantine condition (6.3) for *all* the E_j, we can rewrite (6.9) as

$$
K_{X''} + D'' = \left(k - 3 - \sum_{i=1}^{k} \frac{1}{n_i} \right) H - 3 \sum_{j} \frac{E_j}{m_j} + 3 \sum_{n_i < 0} \frac{D_i}{n_i} + 3 \sum_{m_j < 0} \frac{E_j}{m_j}
$$

$$
= \left(k - 3 - \sum_{i=1}^{k} \frac{1}{n_i} \right) H - 3 \sum_{m_j > 0} \frac{E_j}{m_j} + 3 \sum_{n_i < 0} \frac{D_i}{n_i}. \tag{6.10}
$$

Since the D_s with $n_s < 0$ are orthogonal to $K_{X''} + D''$ and all E_j have zero intersection with $\pi^* H$, we have the following result.

Theorem 6.3 *For a weighted arrangement for which (6.3) holds for all E_j (with positive, negative, and infinite weight) and all D_t with n_t negative or infinite, we have*

$$\left(K_{X''} + D''\right)^2 = \left(k - 3 - \sum_{i=1}^{k} \frac{1}{n_i}\right)^2 - 9\sum_{m_j > 0} \frac{1}{m_j^2} + 9\sum_{n_i < 0} \frac{\sigma_i - 1}{n_i^2}$$

$$= \left(k - 3 - \sum_{i=1}^{k} \frac{1}{n_i}\right)^2 - 9\sum_{j} \frac{1}{m_j^2} + 9\sum_{n_i < 0} \frac{\sigma_i - 1}{n_i^2} + 9\sum_{m_j < 0} \frac{1}{m_j^2}.$$

$$(6.11)$$

Using the tables at the end of Chapter 5, as well as in §5.6.1, and (6.11) of the preceding theorem, one checks that $(K_{X''} + D'')^2$ is positive for all the weighted line arrangements listed in that chapter with $\mathrm{Prop}(Y', C) = 0$. For the complete quadrilateral, we can see this more easily by rewriting (6.11). Let $\mu_0, \ldots, \mu_4$ be five real numbers with $\sum_{i=0}^{4} \mu_i = 2$. Let the $k = 6$ lines of the complete quadrilateral have weights $(1 - \mu_i - \mu_j)^{-1}$, where $1 \leq i < j \leq 4$, and assign the weights $(1 - \mu_i - \mu_0)^{-1}$, where $i = 1, \ldots, 4$, to the four singular points. Then, from Theorem 6.3, we deduce the following formula:

$$\left(K_{X''} + D''\right)^2 / 9 = 1 - \sum_{i=0}^{4} \mu_i^2 + \sum_{i < j, \mu_i + \mu_j > 1} \left(\mu_i + \mu_j - 1\right)^2. \qquad (6.12)$$

It follows easily that this number is positive for the first twenty-seven choices of μ_i for the complete quadrilateral in Table 5.12 of Chapter 5.

Finally, we check that $(K_{X''} + D'') \cdot S > 0$ for all irreducible curves S of X that are not contracted by the map from X to X''. These are the D_i with $n_i > 0$, the E_j with $m_j > 0$, and those curves on X that correspond to the curves on $\mathbb{P}_2$ lifted to X. Assume for simplicity that there are no D_t or E_s of infinite weight. Recall from Chapter 5, §5.4 and §5.6, the definition of the relative proportionality $\mathrm{prop}(\widetilde{D})$ of a smooth curve $\widetilde{D}$ in a smooth compact surface Y. We have

$$\mathrm{prop}(\widetilde{D}) = 2\widetilde{D} \cdot \widetilde{D} - e(\widetilde{D}).$$

Combined with the adjunction formula

$$K_Y \cdot \widetilde{D} = -e(\widetilde{D}) - \widetilde{D} \cdot \widetilde{D},$$

this gives

$$\mathrm{prop}(\widetilde{D}) = K_Y \cdot \widetilde{D} + 3\widetilde{D} \cdot \widetilde{D}.$$

As discussed in Chapter 5, §5.6, the diophantine conditions (6.3) result from the vanishing of the relative proportionalities of the lifts $\widetilde{D}_i$ of D_i and $\widetilde{E}_j$ of E_j on the putative finite cover $Y \to X$ of degree N with ramifications determined, as in Chapter 5, by the weights of the D_i and E_j on X. By the last displayed expression for the relative proportionality, this implies

$$K_Y \cdot \widetilde{D}_i = -3\widetilde{D}_i \cdot \widetilde{D}_i, \qquad K_Y \cdot \widetilde{E}_j = -3\widetilde{E}_j \cdot \widetilde{E}_j.$$

Now, as observed above, the divisor $K_{X''} + D''$ lifts to K_Y if Y exists. Therefore, we can restate these last two equations as

$$(K_{X''} + D'') \cdot \left(\frac{1}{n_i} D_i\right) = -3 \left(\frac{1}{n_i} D_i\right) \cdot \left(\frac{1}{n_i} D_i\right),$$

$$(K_{X''} + D'') \cdot \left(\frac{1}{m_j} E_j\right) = -3 \left(\frac{1}{m_j} E_j\right) \cdot \left(\frac{1}{m_j} E_j\right)$$

so that, for $n_i > 0$ and $m_j > 0$,

$$(K_{X''} + D'') \cdot D_i = -3 \frac{D_i \cdot D_i}{n_i}, \qquad (K_{X''} + D'') \cdot E_j = -3 \frac{E_j \cdot E_j}{m_j}.$$

$$(6.13)$$

Although we have mentioned Y to explain these equations, they are just consequences of the diophantine conditions. If there are D_t or E_s of infinite weight, then (6.13) remains true by working with the quantities $\text{Prop}(\widetilde{D}_i, C)$ and $\text{Prop}(\widetilde{E}_j, C)$ of Chapter 5, §5.6, where C is the union of the elliptic curves on Y coming from the infinite weights. Let S be D_i with $n_i > 0$ or an E_j with $m_j > 0$ in the above equation. Then, as $S \cdot S < 0$, the right-hand sides of the equations in (6.13) are positive. Therefore, if $(K_{X''} + D'') \cdot S \leq 0$, this contradicts (6.13), and so $(K_{X''} + D'') \cdot S > 0$ when S is a D_i with $n_i > 0$ or an E_j with $m_j > 0$.

Now, let S be a curve on X given by the lift of a curve of degree d on $\mathbb{P}_2$. It has intersection number d with every divisor $D_i + \sum_{E_j \cap D_i \neq \phi} E_j$, namely any divisor on X equivalent to the lift to X of a hyperplane H in $\mathbb{P}_2$. For the first twenty-seven cases of the complete quadrilateral considered in Table 5.13 we take, for example, the divisor $D_{12} + D_{03} + D_{04}$. Each of these divisors occurs with a positive coefficient in $K_{X''} + D''$, and so $(K_{X''} + D'') \cdot S > 0$. As we have seen already, in all other cases, except one case for the icosahedral arrangement, the divisor $K_{X''} + D''$ only has positive coefficients, and therefore we have $(K_{X''} + D'') \cdot S > 0$. In the exceptional icosahedral case, we have $(K_{X''} + D'') \cdot S = 0$ only when S corresponds to a curve in $\mathbb{P}_2$ passing through all six quintuple points. The curve S does not correspond to the transforms in X of the fifteen lines of the icosahedral arrangement. Moreover the corresponding curve in $\mathbb{P}_2$ meets each of the fifteen lines of the icosahedral arrangement in the two quintuple points lying on the line. Therefore, the degree of this curve in each of the six quintuple points would have to be $d/2$, with $d \geq 4$ even. Using the Plücker formula, we see that this cannot happen. We remark in passing that the six conics in $\mathbb{P}_2$ passing through five of the six quintuple points lift to one-dimensional ball quotients on X.

We now discuss the weighted arrangements $\mathrm{Ceva}(q)$, $q \geq 3$. These are the only remaining examples where the lines must have equal weights in order to satisfy (6.3). These examples are interesting, though they are special cases of the weighted $\overline{\mathrm{Ceva}}(q)$, $q \geq 3$, arrangements, where the kernel of the matrix R has dimension 4. For $q = 3, 4, 5, 6, 8$, there are altogether seventeen solutions of the diophantine conditions (6.3). A calculation shows that $(K_{X''} + D'')^2 > 0$ in all cases, except $n = -3$, $q = 3$, when $(K_{X''} + D'')^2 = 0$. In this case, formula (6.10) shows that in fact $K_{X''} + D'' = 0$, corresponding to case 29 of the complete quadrilateral. In the other cases, all coefficients of $K_{X''} + D''$ of the form (6.8) are positive except for the E_j with $m_j = 1$, which correspond to the three q-tuple points, $q \geq 4$. A curve S of degree d coming from the projective plane, with at least two q-tuple points, must intersect the transforms in X of the lines going through these q-tuple points. It is possible this intersection meets the blown-up triple points. For the arrangements $\overline{\mathrm{Ceva}}(q)$, $q \geq 2$, recall that twenty-two weighted arrangements come directly from $\overline{\mathrm{Ceva}}(1)$ by using a covering $\mathbb{P}_2 \to \mathbb{P}_2$ of degree q^2, as in (5.48). These correspond to twenty-two of the twenty-seven cases in Table 5.12, namely all of them except cases 8, 10, 11, 24, and 27. In these cases, $(K_{X''} + D'')^2$ for $\overline{\mathrm{Ceva}}(q)$ equals q^2 times that for the corresponding $\overline{\mathrm{Ceva}}(1)$, and the $K_{X''} + D''$ of $\overline{\mathrm{Ceva}}(1)$ simply lifts to that of $\overline{\mathrm{Ceva}}(q)$.

We have therefore checked that for the weighted line arrangements of Chapter 5, whose weights satisfy all the diophantine conditions (6.3), the divisor $K_{X''} + D''$ is effective, has $(K_{X''} + D'')^2 > 0$, and $(K_{X''} + D'') \cdot S'' > 0$ for every irreducible curve S'' on X''. Therefore, $K_{X''} + D''$ is ample and defines a map from X to a surface isomorphic to X'' embedded in a projective space.

6.6 LOG-TERMINAL SINGULARITIES AND LCS

In this section we identify the log-canonical singularities for the log-canonical model (X'', D'') of the pair (X, D) coming from a weighted line arrangement. This helps to relate Theorems 6.1 and 6.2 of §6.1 to the particular application that results in Theorem 6.4. Weighted line arrangements give rise to orbifold singularities whose nature and contribution to the Miyaoka-Yau inequality can be identified independently of Definition 6.2. It is nonetheless instructive to work with this definition for these examples.

With the notation of §6.1, specialized to our situation as described in §6.3, let (V, D, p) be a germ of the normal surface pair (X'', D'') coming from a weighted line arrangement satisfying the diophantine conditions (6.3) and let $(\widetilde{V}, \widetilde{D}, \widetilde{E})$ be a resolution of (V, D, p). When $\widetilde{E} = E_s$ is an exceptional divisor with negative or infinite weight m_s, we have (see Chapter 3)

$$
\begin{aligned}
\pi^*(K_V + D) = \quad & K_{\widetilde{V}} + \widetilde{D} + a\widetilde{E} \\
= \; & \pi^* K_V + \widetilde{E} + \widetilde{D} + a\widetilde{E}.
\end{aligned}
$$

By the adjunction formula,

$$
K_{\widetilde{V}} \cdot \widetilde{E} = -\widetilde{E} \cdot \widetilde{E} - e(\widetilde{E}) = 1 - 2 = -1.
$$

Therefore, if r lines with weights $n_i \geq 1$ intersect at the point p_s which is blown up to E_s, we have

$$
\left(K_{\widetilde{V}} + \widetilde{D} + a\widetilde{E} \right) \cdot \widetilde{E} = -1 + \sum_{i=1}^{r} \left(1 - \frac{1}{n_i} \right) - a = 0,
$$

so that

$$
\begin{aligned}
a = \quad & r - 1 - \sum_{i=1}^{r} \frac{1}{n_i} \\
= \; & 1 + \left(r - 2 - \sum_{i=1}^{r} \frac{1}{n_i} \right) \\
= \; & 1 + \frac{2}{m_s}
\end{aligned}
$$

by the first diophantine condition in (6.3). If $m_s < 0$, then $a < 1$, and if $m_s = \infty$, then $a = 1$. Therefore, by Definition 6.2, if $m_s < 0$ or ∞, then (V, D, p_s) is a *log-canonical singularity*. It is in addition *log-terminal* when $m_s < 0$, and LCS when $m_s = \infty$. By convention ([84], p. 342), one can consider a regular point p, with $D = \phi$, or a quasi-regular point p, with $D = (1 - \frac{1}{b_1})D_1$ and $2 \leq b_1 < \infty$, of a germ (V, D, p) to be log-terminal: indeed, this fits with the definition. In the latter case, it is uniformized by the former via the map, in local coordinates, $(z_1, z_2) \to (z_1^{b_1}, z_2)$ with covering group $(\mathbb{Z}/b_1\mathbb{Z}) \times \{1\}$.

If we blow down to a point p_t a proper transform D_t of a line of the arrangement, with weight n_t negative or infinite, then there will be τ_t lines passing through p_t. The E_j with $E_j \cap D_t \neq \phi$ will have positive weights, as will the D_ℓ with $D_\ell \cap D_t \neq \phi$. By the adjunction formula, we have, using the above notation,

$$
K_{\widetilde{V}} \cdot D_t = -D_t \cdot D_t - e(D_t) = \sigma_t - 3,
$$

as $D_t \cdot D_t = 1 - \sigma_t$ and $e(D_t) = 2$. We have

$$
\pi^*(K_V + D) = \quad K_{\widetilde{V}} + \widetilde{D} + aD_t,
$$

where a is determined by

$$\left(K_{\widetilde{V}} + \widetilde{D} + a D_t\right) \cdot D_t = 0.$$

Therefore,

$$\sigma_t - 3 + \sum_{D_j \cap D_t \neq \phi} \left(1 - \frac{1}{m_j}\right) + \sum_{D_\ell \cap D_t \neq \phi} \left(1 - \frac{1}{n_\ell}\right) + a(1 - \sigma_t) = 0,$$

which implies that

$$(1-a)(\sigma_t - 1) + \tau_t - 2 - \sum_{D_j \cap D_t \neq \phi} \frac{1}{m_j} - \sum_{D_\ell \cap D_t \neq \phi} \frac{1}{n_\ell}$$
$$= (1-a)(\sigma_t - 1) + \frac{2(\sigma_t - 1)}{n_t} = 0. \tag{6.14}$$

In the last equality, we have used the second diophantine condition in (6.3). Therefore, $a = 1 + \frac{2}{n_t}$, as $\sigma_t \neq 1$, so that when $n_t < 0$ we have $a < 1$, and when $n_t = \infty$ we have $a = 1$. Therefore, we have a log-terminal singularity when $n_t < 0$ and an LCS when $n_t = \infty$. Notice that we could have combined the above calculations for the E_s and the D_t into a single one.

When $n_t < 0$, only the cases $\sigma_t = 2, 3$ can occur. This is because the second diophantine condition in (6.3) is the same as that of Chapter 5, (5.32), whose only solutions occur for $r = 2, 3$, where r corresponds to $\tau_t \geq \sigma_t \geq 2$. When $n_t = \infty$, by Chapter 5, (5.33), as well as the cases $\sigma_t = 2, 3$, we also have one case when $\tau_t = 4$. It corresponds to a contraction of D_t to a singular point p_t with all lines passing through p_t having weight 2. This occurs for the configuration Ceva(3) with $n = \infty$ discussed in Chapter 5, §5.6, which corresponds to entry 23 of Table 5.12.

Finally, we observe that a regular transverse intersection point of two weighted divisors $D_i = D_i''$, of weight $1 \leq b_i < \infty$, and $D_j = D_j''$, of weight $1 \leq b_j < \infty$, is a log-terminal singularity of X''. Let

$$D = \left(1 - \frac{1}{b_i}\right) D_i + \left(1 - \frac{1}{b_j}\right) D_j,$$

and blow up the point $p = D_i \cap D_j$. Let $\widetilde{E} = E$ be the resulting exceptional divisor. Then $(\widetilde{V}, \widetilde{D}, E)$ is the resolution of the normal surface germ (V, D, p) and $\Delta = aE$, where

$$\begin{aligned}
\left(K_{\widetilde{V}} + \widetilde{D} + \Delta\right) \cdot E &= K_{\widetilde{V}} \cdot E + \widetilde{D} \cdot E + aE \cdot E \\
&= -1 + \left(1 - \frac{1}{b_i}\right) + \left(1 - \frac{1}{b_j}\right) - a \\
&= 1 - a - \left(\frac{1}{b_i} + \frac{1}{b_j}\right) = 0.
\end{aligned}$$

In the second line of the above formula, we have used the adjunction formula, and we have

$$1 - a = \left(\frac{1}{b_i} + \frac{1}{b_j} \right) > 0,$$

so that $a < 1$, and p is therefore a log-terminal singularity of X''.

Log-canonical surface singularities have been classified [114] (see also [84], Theorem 3.1). Every germ (V, D, p) for which p is regular and $D = \phi$ or $D = (1 - \frac{1}{b})D$ with $b < \infty$ is log-terminal. As mentioned already, in the latter case it is uniformized by $(z_1, z_2) \to (z_1^b, z_2)$ with covering group $\mathbb{Z}/b\mathbb{Z} \times \{\text{Id}\}$. An isolated singularity (V, ϕ, p) is log-terminal if and only if it is a quotient singularity. The exceptional set in the minimal resolution is a configuration of complex projective lines associated to, and determined by, ADE Dynkin diagrams. It is uniformized by $\mathbb{C}^2$, and the covering transformation group is a finite subgroup of U(2) free of reflections. As X'' is smooth away from the support of D'', these singularities do not occur in our case, but they did arise in the discussion of the Miyaoka-Yau inequality in Chapter 4, §4.2. Finally, the *log-terminal singularities* with $D \neq \phi$ are quotients $\mathbb{C}^2/G$, where G is a finite subgroup of U(2), which is an extension of a unitary reflection group. The analytic types of the singularities are in 1-1 correspondence with the conjugacy classes of the finite subgroups of U(2), the quasi-regular log-terminal singularities being in 1-1 correspondence with the conjugacy classes of the unitary reflections groups. When (X'', D'') is a (possibly compactified) ball quotient $\overline{\Gamma \backslash B_2}$, as in the Question of §6.4, the points in X'' corresponding to the E_s with $m_s < 0$ and to the D_t with $n_t < 0$ are quotient singularities with respect to finite isotropy subgroups of Γ, and these isotropy groups can be determined from the weight data of (X, D). The points corresponding to the E_s with $m_s = \infty$ and D_t with $n_t = \infty$ come from cusps corresponding to infinite isotropy groups of points at the boundary of the ball. We discuss the isotropy subgroups of Γ in more detail after the proof of Theorem 6.4.

6.7 EXISTENCE THEOREM FOR LINE ARRANGEMENTS

In this section, we state and discuss the proof of the main existence theorem, namely Theorem 6.4 below, ensuring that the line arrangements of Chapter 5 whose weights satisfy the diophantine conditions (6.3) have finite covers that are free ball quotients. The proof of Theorem 6.4 proceeds by first addressing the Question of §6.4, which asks whether, under the diophantine conditions (6.3) for *all* weights, i.e., those of all proper transforms of the lines of an arrangement and all exceptional divisors coming from blown-up points, the surface X'' is an appropriately ramified compactified ball quotient. If a finite cover Y of X as described in Chapter 5, §5.6, exists, then these diophantine conditions are equivalent to $\mathrm{Prop}(Y', C) = 0$, where Y' is the surface obtained from Y by blowing down the rational curves covering the D_s and E_t of negative weight and C is the image on Y' of the elliptic curves on Y covering the D_s and E_t of infinite weight. The surface $Y' \setminus C$ is then the ball quotient whose existence we seek. Our key references are [83], [84], as well as the series of papers by R. Kobayashi, [79], [80], [81], [82], leading to these two references. Those elements of the discussion of the proof that are not justified, or otherwise referenced, have their origin in these key references, which we will not continue to cite on every relevant occasion. What we add to the discussion of those two texts is some filling out of basics and, more importantly, the application of their discussion to our particular context. Recall that (X'', D'') is called the log-canonical model of (X, D). It plays the role, for D nontrivial, of the canonical model of X discussed in Chapter 4. We also refer to Chapter 4, §4.3, for the basic definitions relating to Kähler and Kähler-Einstein metrics used in what follows.

In the course of the proof of Theorem 6.4, we shall derive, under the assumption that $\overline{\kappa}(X, D) = 2$, the following Miyaoka-Yau inequality for (X_0'', D_0''), which is an equality when all the diophantine conditions in (6.3) are fulfilled, in which case (X_0'', D_0'') is a ball quotient as in the Question of §6.4. This is a special case of the Miyaoka-Yau inequality for orbifolds in Theorem 6.2, §6.1, and is given by

$$\frac{1}{3}\left(K_{X''} + D''\right)^2 \le e(X_0'') + \sum_{n_i > 0}\left(\frac{1}{n_i} - 1\right)\left(e(D_{0,i}'') - d_i\right)$$

$$+ \sum_{m_j > 0}\left(\frac{1}{m_j} - 1\right)\left(e(E_{0,j}'') - e_j\right) + \sum_{P}\left(\frac{1}{|G_P|} - 1\right),$$

where P, with local uniformizing group G_P, runs over the intersection points of D_0'', and where d_i is the number of intersection points of D'' above $D_{0,i}''$ and e_j is the number of intersection points of D'' above $E_{0,j}''$. Recall that $D_0'' = D'' \cap X_0''$, where D'' is the image of D in X''. Moreover, D_i'' is the image of D_i, with $n_i > 0$ in X'', and $D_{0,i}'' = D_i'' \cap X_0''$; similarly for $E_{0,j}''$ with $m_j > 0$.

Theorem 6.4 *Consider a weighted line arrangement with weights n_i for the lines and m_j for the singular points. Assume that the diophantine conditions (6.3) are satisfied for all weights. Let X be the blow-up of $\mathbb{P}_2$ at the singular points of the arrangement. Then, consider in X the proper transforms D_i of the lines and the blow-ups E_j of the singular points of the arrangement. Assume that the E_s, with $m_s < 0$ or infinite, and the D_t, with $n_t < 0$ or infinite, are disjoint.*

Let X'' be the (possibly singular) surface obtained by contracting them. For the given data, we have defined $K_{X''} + D''$. Suppose $K_{X''} + D''$ can be represented by an effective divisor, that $(K_{X''} + D'')^2 > 0$, and that $(K_{X''} + D'') \cdot S > 0$ for any curve S on X'' determined by a curve in $\mathbb{P}_2$. Then, there is a good covering Y of X with ramifications n_i' and m_j' along the D_i and the E_j, where $n_i' = |n_i|$, $m_j' = |m_j|$ for n_i, $m_j \neq \infty$, and n_i', m_j' are arbitrary positive integers if $n_i = \infty$ or $m_j = \infty$. We have a union of exceptional curves over the E_s with $m_s < 0$, and over the D_t with $n_t < 0$, and a union C' of elliptic curves over the E_s with $m_s = \infty$, and over the D_t with $n_t = \infty$. Blowing down the exceptional curves, we obtain a smooth surface Y'. Then, $Y' \setminus C$, where C is the image of C' in Y, is a ball quotient

$$Y' \setminus C = \Gamma' \backslash B_2$$

for a discrete subgroup Γ' of $\mathrm{Aut}(B_2) \simeq \mathrm{PU}(1, 2)$ operating freely. Moreover, $Y' \setminus C$ has finite volume. The elliptic curves correspond to the cusps of Γ': they originate in 1-point compactifications, with the resulting singularity blown up to an elliptic curve.

Proof. By the discussion of the ampleness criteria for $K_{X''} + D''$ in §6.5, the assumptions of the theorem imply that, for some integer $m > 0$, the divisor $m(K_{X''} + D'')$ is very ample and the irreducible components of its support have nonnegative integer coefficients. Moreover, the only divisors on X that have zero intersection with $K_{X''} + D''$ are the divisors D_t and E_s with negative or infinite weight. Therefore, the line bundle associated to $m(K_{X''} + D'')$ has a finite number of global sections $r_0, \dots, r_N$ with the property that, for any $x \in X$, they do not all vanish at x and they define a holomorphic map

$$\varphi_m : X \to \mathbb{P}_N$$

given by

$$x \mapsto [r_0(x) : r_1(x) : \dots : r_N(x)],$$

which is biholomorphic on the complement of the divisors D_t and E_s of negative and infinite weight, and constant on those same divisors. (When D is empty, this is the m-canonical map of §4.1.) The image $\varphi_m(X)$ of X in $\mathbb{P}_N$ is a projective algebraic variety isomorphic to X'', which possibly has a finite number of quotient singularities. Recall that, in our case, these singularities will occur at the blow-downs of the D_t with negative or infinite weight and self-intersection $D_t^2 \leq -2$. When D_t has negative weight, we have $D_t^2 = -1$ or -2, and when D_t has infinite weight, we have $D_t^2 = -1, -2$, or -3.

Let $\mu : X \to X''$ be the map that contracts the divisors D_t and E_s on X with negative and infinite weight. Then the map φ_m is the composition $\rho_m \circ \mu$ of μ

with an embedding ρ_m of X'' into $\mathbb{P}_N$, called a *log-canonical mapping*:

$$\rho_m : X'' \to \mathbb{P}_N(\mathbb{C}).$$

Projective space is endowed with a Kähler-Einstein metric FS, called the *Fubini-Study metric*, mentioned in §4.3, and given in the coordinate patch $z_i \neq 0$, for homogeneous coordinates $[z_0 : z_1 : \ldots : z_N]$ of $\mathbb{P}_N$, by the Hermitian matrix

$$FS_{k\bar{\ell}} = \frac{\left(1 + \sum_k |w_k|^2\right)\delta_{k\ell} - \overline{w}_k w_\ell}{\left(1 + \sum_k |w_k|^2\right)^2} = \partial^2 \log\left(1 + \sum_k |w_k|^2\right)/\partial w_k \partial \overline{w}_\ell,$$

where $w_k = z_k/z_i$, $k = 1, \ldots, N$, are the corresponding affine coordinates. Denoting by ω_{FS} the $(1,1)$-form of FS, we have, in terms of the homogeneous coordinates,

$$\omega_{FS} = \sqrt{-1}\partial\overline{\partial}\log\|z\|^2$$

$$= \sqrt{-1}\partial\overline{\partial}\log(|z_0|^2 + \ldots + |z_N|^2).$$

Define ω_m to be the pullback of ω_{FS} to X. Then, for $x \in X$,

$$\omega_0 = \frac{1}{m}\omega_m = \frac{1}{m}((\rho_m \circ \mu)^*(\omega_{FS}))(x)$$

is semipositive on X: it is positive definite outside the curves contracted by ρ_m, and $\iota_{\widetilde{E}}^* \omega_0 = 0$. Here $\widetilde{E}$ is the union of the divisors E_s and D_t with negative or infinite weight and $\iota_{\widetilde{E}}$ is the inclusion of $\widetilde{E}$ in X. Let (u_1, u_2) be local coordinates on an open set U of X such that the map $(\rho_m \circ \mu)$ on U has image in the affine subset of $\mathbb{P}_N$ given by $z_0 \neq 0$. Then, for $u = (u_1, u_2) \in U$, we can give an explicit formula for $\omega_0(u)$ as follows. We have

$$\psi_m(u) := (\rho_m \circ \mu)(u) = [1 : s_1(u) : \ldots : s_N(u)]$$

where $s_j = r_j/r_0$, $j = 1, \ldots, N$. Let $\partial\overline{\partial}$ be the operator

$$\partial\overline{\partial} = \sum_{k,\ell=1}^{2} \frac{\partial}{\partial u_k}\frac{\partial}{\partial \overline{u}_\ell} du_k \wedge d\overline{u}_\ell.$$

Then, by the definition of the pullback of a $(1, 1)$-form, for $u \in U$ we have,

$$\omega_0(u) = -\frac{1}{m}\sqrt{-1}\partial\bar{\partial}\log\left(1 + \sum_i |s_i|^2\right)$$

$$= -\frac{1}{m}\sqrt{-1}\sum_{k,\ell=1}^{2} \frac{\partial}{\partial u_k}\frac{\partial}{\partial \bar{u}_\ell}\log\left(1 + \sum_i |s_i(u)|^2\right) du_k \wedge d\bar{u}_\ell$$

$$= -\frac{1}{m}\sqrt{-1}\sum_{k,\ell=1}^{2}\sum_{p,q=1}^{N} \frac{\partial s_p}{\partial u_k}\frac{\partial \bar{s}_q}{\partial \bar{u}_\ell}\frac{\partial}{\partial s_p}\frac{\partial}{\partial \bar{s}_q}\log\left(1 + \sum_i |s_i|^2\right) du_k \wedge d\bar{u}_\ell$$

$$= -\frac{1}{m}\sqrt{-1}\sum_{k,\ell=1}^{2}\sum_{p,q=1}^{N} \frac{(1 + \sum_i |s_i(u)|^2)\delta_{pq} - \overline{s_p(u)}s_q(u)}{(1 + \sum_i |s_i(u)|^2)^2}\frac{\partial s_p}{\partial u_k}\frac{\partial \bar{s}_q}{\partial \bar{u}_\ell} du_k \wedge d\bar{u}_\ell$$

$$= \frac{1}{m}\sum_{p,q=1}^{N}\left(\frac{(1 + \sum_i |s_i(u)|^2)\delta_{pq} - \overline{s_p(u)}s_q(u)}{(1 + \sum_i |s_i(u)|^2)^2}\right)\left(-\sqrt{-1}\partial\bar{\partial}s_p\bar{s}_q(u)\right).$$

The map μ has domain a minimal good resolution

$$\mu : \left(X, \tilde{D}, \tilde{E}\right) \rightarrow \left(X'', D''\right)$$

of (X'', D''), where $\tilde{D}$ is the proper transform of D'' and $\tilde{E} = \sum_\alpha E_\alpha$ is the exceptional set. Here, we have

$$\tilde{D} = \sum_{n_i \geq 2}\left(1 - \frac{1}{n_i}\right) D_i + \sum_{m_j \geq 2}\left(1 - \frac{1}{m_j}\right) E_j,$$

and

$$\tilde{E} = \sum_s E_s + \sum_t D_t,$$

where s indexes the infinite and negative weights m_s, and t indexes the infinite and negative weights n_t. Recall that the D_i and D_t are the proper transforms of the lines of the arrangement, and the E_j and E_s are the blow-ups of its singular points. Denoting as before $\mu^*(K_{X''} + D'')$ simply by $K_{X''} + D''$, we have

$$K_{X''} + D'' = K_X + \tilde{D} + \sum_\alpha a_\alpha E_\alpha,$$

with $a_\alpha \leq 1$. By the discussion of §6.6,

$$a_s = 1 + \frac{2}{m_s}, \qquad a_t = 1 + \frac{2}{n_t},$$

so that $a_s, a_t < 1$ when the corresponding weights are negative, whereas $a_s = a_t = 1$ when the corresponding weights are infinite.

Alternatively, a short computation shows that

$$K_{X''} + D'' = \pi^* K(\mathbb{P}_2) + \sum_{i=1}^k \left(1 - \frac{1}{n_i}\right) D_i + \sum_j \left(2 - \frac{1}{m_j}\right) E_j$$

$$+ \sum_t \left(a_t - 1 + \frac{1}{n_t}\right) D_t + \sum_s \left(a_s - 1 + \frac{1}{m_s}\right) E_s,$$

where the ranges of s and t are as before. If we now take into account the diophantine conditions, the above formula must coincide with that in (6.7), so that we recover the values of a_s and a_t, namely

$$a_s = 1 + \frac{2}{m_s}, \qquad a_t = 1 + \frac{2}{n_t},$$

and the fact that $a_s, a_t < 1$ when the corresponding weights are negative, and $a_s = a_t = 1$ when the corresponding weights are infinite.

We now change notation so as to be closer to that of our central references [83], [84]. Let b_k run over the $n_i > 0$ and $m_j > 0$, and let σ_k be a holomorphic section of the line bundle associated to the divisor D_i or E_j with weight b_k. Let b_ℓ run over the $n_t < 0$ and $m_s < 0$, and let σ_ℓ be a holomorphic section of the line bundle associated to the divisor D_t or E_s of weight b_ℓ. Finally, let b_m run over the n_t and m_s infinite, and let σ_m be a holomorphic section of the line bundle D_t or E_s associated to the divisor of weight b_m. Let $\| \cdot \|$ denote a Hermitian metric of a line bundle. By the discussion of Chern classes in §1.4, there exists a smooth volume form Ω on X such that $\omega_0 = -\mathrm{Ric}(\Phi)$ (a typo in [84], p. 353, displays this with the wrong sign as $\omega_0 = \mathrm{Ric}(\Phi)$), where Φ is the singular volume form

$$\Phi = \frac{\Omega}{\prod_{b_k > 0} \|\sigma_k\|^{2(1-\frac{1}{b_k})} \prod_{b_\ell < 0} \|\sigma_\ell\|^{2(1+\frac{2}{b_\ell})} \prod_{b_m = \infty} \|\sigma_m\|^2}.$$

We now check that Φ is the lift to X of a $b - C^\infty$ volume form for the orbifold (X_0'', D_0''). In what follows, B will always denote the 2-ball in $\mathbb{C}^2$, centered at the origin $(0, 0)$, and of suitably small radius. Let $\check{D}_1''$ with weight $b_1 > 0$ and $\check{D}_2''$ with weight $b_2 > 0$ be two divisors in $\text{Supp}(D'')$ that intersect transversally on X'' at a point p_{12}. Here $\check{D}_1''$ can be a D_i'' or E_j''. At a point p_i of $\check{D}_i$, $i = 1, 2$, with $p_i \neq p_{12}$, the orbifold (X'', D'') has a local uniformization of the form $(B, (\mathbb{Z}/b_i\mathbb{Z}) \times \{1\}, \pi_i)$ where

$$\pi_i(z_1, z_2) = (z_1^{b_i}, z_2) = (u, v)$$

for local coordinates (u, v) on X'' centered at p_i such that $u = 0$ is the local equation for $\check{D}_i''$. On B, the standard Euclidean volume form

$$v_e = dz_1 \wedge dz_2 \wedge d\bar{z}_1 \wedge d\bar{z}_2$$

is $((\mathbb{Z}/b_i\mathbb{Z}) \times \{1\})$-invariant. In terms of (u, v) it becomes

$$\frac{1}{b_i^2}|u|^{-2(1-\frac{1}{b_i})}du \wedge dv \wedge d\bar{u} \wedge d\bar{v}.$$

At p_{12}, a local uniformization of (X'', D'') is $(B, (\mathbb{Z}/\mathbb{Z}b_1) \times (\mathbb{Z}/\mathbb{Z}b_2), \pi_{12})$ where $\pi_{12}(z_1, z_2) = (z_1^{b_1}, z_2^{b_2}) = (u, v)$ and (u, v) are local coordinates on X'' centered at p_{12}. Here u is the normal to D_1'' and v is the normal to D_2''. The volume form v_e on B is $(\mathbb{Z}/\mathbb{Z}b_1) \times (\mathbb{Z}/\mathbb{Z}b_2)$-invariant, and in terms of (u, v) is given by

$$\frac{1}{b_1^2 b_2^2}|u|^{-2(1-\frac{1}{b_1})}|v|^{-2(1-\frac{1}{b_2})}du \wedge dv \wedge d\bar{u} \wedge d\bar{v}.$$

Therefore, in a suitable neighborhood U_p of a point p on D_i'' or E_j'', which is not the image of the contraction of a divisor in $\text{Supp}(D)$ of negative or infinite weight, the volume form Φ is locally the lift to X of $h_p v_p$ where v_p is the push-down to U_p of v_e and h_p is a smooth function on U_p. Therefore Φ is locally the lift to X of a $b - C^\infty$-volume form on (X'', D'').

Now, let $p_\ell \in X''$ be the image of a divisor $\check{D}_\ell$ on X of negative weight b_ℓ (in other words, $\check{D}_\ell$ is a D_t with $n_t < 0$ or an E_s with $m_s < 0$). Let $\mu_\ell : (\check{U}_\ell, \check{D}_\ell) \to (U_\ell, p_\ell)$, where $\check{U}_\ell$ is an open subset of X, be a resolution of (U_ℓ, p_ℓ) such that the closure of $\check{U}_\ell$ does not intersect any other divisor D_t or E_s of negative or infinite weight. Let (B, G_ℓ, π_ℓ) be a local orbifold uniformization of (X'', D'') at p_ℓ. Then G_ℓ is a finite subgroup of $U(2)$, so that v_e is G_ℓ-invariant. Therefore, the singular volume form $v_\ell = (\pi_\ell)_*(v_e)$ on U_ℓ, induced by v_e, is well defined. Let $\check{v}_\ell = v_\ell \circ \mu_\ell$ on $\check{U}_\ell$. Then, from our observations above, the singular volume form Φ, restricted to $\check{U}_\ell$, is of the form $\check{h}_\ell \check{v}_\ell$ for a function $\check{h}_\ell$ that is smooth on $\check{U}_\ell \setminus \check{D}_\ell \simeq U_\ell \setminus \{p_\ell\}$. Let

$h_\ell = \check{h}_\ell|_{U_\ell \setminus \{p_\ell\}}$. Then, on $B \setminus \{(0,0)\}$, we have

$$\pi_\ell^*(h_\ell v_\ell) = (h_\ell \circ \pi_\ell) v_\ell,$$

and $h_\ell \circ \pi_\ell$ is smooth on $B \setminus \{(0,0)\}$. On the other hand, consider the $(1,1)$-Kähler form on B given by $\frac{1}{m}((\rho_m \circ \pi_\ell)^*(\omega_{FS}))(z)$, $z \in B$, and let h'_ℓ be a smooth Kähler potential for this form on B. Then, as we have $\omega_0 = -\mathrm{Ric}(\Phi)$, it follows that $\partial \bar{\partial} \log(h_\ell \circ \pi_\ell) = \partial \bar{\partial} h'_\ell$, where

$$\partial \bar{\partial} = \sum_{i=1,2} \frac{\partial}{\partial z_1} \frac{\partial}{\partial \bar{z}_2} dz_1 \wedge d\bar{z}_2,$$

for coordinates (z_1, z_2) on B. Therefore, $h_\ell \circ \pi_\ell$ can be extended to a smooth function on all of B by the theorem of removable singularities for harmonic functions (see, for example, [9], Theorem 2.3). Therefore, Φ is locally the lift of a $b - C^\infty$ volume form on (X'', D'').

Next, we consider the orbifold (X''_0, D''_0) in a suitable neighborhood of a deleted point p, where that point is the image in X'' of a contracted divisor $\check{D}_s$ on X that is either an E_s or a D_t with infinite weight. For a suitable open set $U_p \subset X''$ containing p, the local orbifold uniformization of (X''_0, D''_0) at $U_p \setminus \{p\}$ is of the form (S_L, Γ, π), for some $L > 0$, where

$$S_L = \{(u : v : 1) \in \mathbb{P}_2 : \Im(u) - |v|^2 > L\}$$

is a subset of

$$S = \{(u : v : 1) \in \mathbb{P}^2 : \Im(u) - |v|^2 > 0\}$$

and Γ is a discrete subgroup of the parabolic subgroup of $\mathrm{Aut}(S)$ fixing $(1 : 0 : 0)$. The domain S is isomorphic to B_2. We can argue in exactly the same way as for the case of negative weight that Φ is locally the lift to $X \setminus \check{D}_s$ of a $b - C^\infty$ volume form on X''_0: indeed we cannot, and do not need to, extend the $b - C^\infty$ volume form to the deleted point p. Ultimately, the b-metric that we set up will not be defined at p, but it will be complete on X''_0. Moreover, the end corresponding to p will have finite volume and is pseudo-concave in the sense that the Kähler potential tends to $-\infty$ at p (for more details, see [84], pp. 354–355).

We have shown that $\mu_*(\Phi)$, restricted to X''_0, is a $b - C^\infty$ volume form for the orbifold (X''_0, D''_0). We now modify Φ by a function $\mu^*(g)$ where g is a $b - C^\infty$ function on the orbifold (X''_0, D''_0), chosen so that $\mu_*(\mu^*(g)\Phi)$ lifts locally to a form k in the local uniformizations such that $-\mathrm{Ric}(k)$ is the ball metric, and not the flat Euclidean metric. Recall that the ball metric has Kähler potential $\frac{1}{2} \log K$, where

$$K = (1 - |z_1|^2 - |z_2|^2)^{-2}.$$

For $x \in X_0''$, we treat the different local uniformizations by a small ball B centered at the origin in a similar way as in the above discussion, where we showed that $\mu_*(\Phi)|_{X_0''}$ is a $b - C^\infty$ volume form for the orbifold (X_0'', D_0''). Along the way, we incorporate the factors b_i^{-2} "missing" from our discussion that Φ is locally the push-down of the standard Euclidean volume form on B: in fact, they are incorporated so that the limit as $b_i \to \infty$ exists and equals the factor needed for a divisor of infinite weight (see [84], pp. 355–356). In our case, the divisors of infinite weight on X and X' are all blown down to form X''. Namely, in a neighborhood of a point $p \in D_{i,0}''$, where $D_{i,0}''$ has weight $b_i > 0$, and p is not an intersection point with another divisor on $D_{i,0}''$, we have, with the same notation as before,

$$\mu^*(g) \simeq b_i^{-2}(1 - \|\sigma_i\|^{2/b_i})^{-2}.$$

In a neighborhood of the intersection of $D_{i,0}''$ of weight b_i and $D_{j,0}''$ of weight b_j, which intersect transversally, we have

$$\mu^*(g) \simeq b_i^{-2} b_j^{-2}(1 - \|\sigma_i\|^{2/b_i})^{-2}(1 - \|\sigma_j\|^{2/b_j})^{-2}.$$

In a suitable neighborhood U_s of a point p_s, which is the image on X'' of the blow-down of $\check{D}_s$, which is an E_s or D_t with negative weight, we have the resolution $(\check{U}_s, \check{D}_s)$ of (U_s, p_s) on X. We have

$$\mu^*(g|_{U_s \setminus \{p_s\}}) = \mu^*(g)|_{\check{U}_s \setminus \check{D}_s} \simeq \prod_i b_i^{-2}(1 - \|\sigma_i\|^{2/b_i})^{-2},$$

the product being over the divisors of weight $b_i > 0$ intersecting $\check{D}_s$ on X.

Let p_s be the image in X'' of the blow-down of $\check{D}_s$, which is an E_s or D_t with infinite weight. As a (possible) singularity of X'', the point p_s is resolved on X by the rational curve $\check{D}_s$. However, as observed above, the local orbifold chart of (X_0'', D_0'') near the deleted point p_s is of the form (S_L, Γ, π), where Γ is a discrete subgroup of the group of automorphisms of S fixing $(1 : 0 : 0)$. It is therefore a "simple elliptic singularity" of $\Gamma \backslash S \cup \{p_s\}$ (see [84], p. 344), as it is resolved by an elliptic curve C with negative self-intersection (see Chapter 4, §4.3, Chapter 5, §5.4). The Kähler potential for the ball metric in the description of B_2 as S is $K(u, v) = \log(\Im(u) - |v|^2)^{-1}$, $(u : v : 1) \in \mathbb{P}_2$. Since the function $F(u, v) = (\Im(u) - |v|^2)$ is invariant under the action of Γ, it projects down to a function f_{p_s} on $U_s \setminus \{p_s\}$, and $-\sqrt{-1}\partial\bar{\partial} \log(f_{p_s})$ is the ball metric near $\{p_s\}$. Moreover, it is complete toward the end corresponding to p_s, and the end has finite volume.

As already discussed, by a b-metric, or orbifold metric, we mean a usual metric outside the ramification locus of the orbifold and, in a neighborhood of a point on the ramification locus with local uniformization (U, G), a metric induced by a G-invariant metric on U. (Strictly speaking, we also need a "gluing" condition, namely that under an injection $U' \to U$ of charts upstairs, the pullback is the corresponding metric on U'.) Such a metric induces a metric on the underlying space of the orbifold, but with possible singularities along the orbifold locus.

Returning to our construction, we obtain in this way a singular volume form Ψ such that $\mu_*(\Psi)$ is $b - C^\infty$ on the orbifold (X_0'', D_0'') and lifts, in the local uniformizations by a small 2-ball about the origin, to a volume form whose negative Ricci form is the $(1, 1)$-form of the ball metric on B. That is, we have:

(i) $\mu_*(\Psi)$ is $b - C^\infty$ and has negative Ricci form, so that $\omega = -\mathrm{Ric}(\Psi)$ defines a Kähler b-metric $\mu_*(\omega)$ on (X_0'', D_0''). Notice that

$$-\mathrm{Ric}(\Psi) = -\mathrm{Ric}(\Phi) + \sqrt{-1}\partial\bar\partial \log(g).$$

Recall that, by construction, the $(1, 1)$-form $\omega_0 = -\mathrm{Ric}(\Phi)$ is semi-positive definite and positive definite outside the curves contracted by μ. Moreover, we verified that it descends to a $b - C^\infty$ $(1, 1)$-form on (X_0'', D_0''). The function g can be constructed in such a way that $\sqrt{-1}\partial\bar\partial \log(g)$ only serves to make $-\mathrm{Ric}(\Psi)$ more positive, since it is constructed to lift to the $(1, 1)$-form of the ball metrics in the local uniformizations. We also have, by construction, that

(ii) the Kähler b-metric ω is complete on X_0'', and, as Ψ is $b - C^\infty$, we have

(iii) the function $f = \log(\Psi/\omega^2)$ is $b - C^\infty$, since Ψ and ω^2 are two volume forms that are $b - C^\infty$. Moreover, it is bounded on X_0'', since the behavior at the ends is controlled. In particular, $\log(g)$ is strictly plurisubharmonic (in that $\sqrt{-1}\partial\bar\partial \log(g)$ is a positive $(1, 1)$-form), and goes to $-\infty$ at the ends of X_0''. For this last property and more details on the whole construction of Ψ, see [80], [81], and [84].

We now introduce the notion of quasi-coordinate system as in [81]. Let V be a domain in $\mathbb{C}^m$ and X an m-dimensional complex manifold. Let ϕ be a holomorphic map of V into X that is everywhere of maximal rank, that is, the rank of the Jacobian map is maximal, so of rank m. In particular, the map ϕ is locally invertible. We call (V, ϕ) a quasi-coordinate system and ϕ a quasi-coordinate map.

Consider (X'', D'', ω), with (X'', D'') and $\omega = -\mathrm{Ric}(\Psi)$, as above. Then, using arguments similar to those of [81], Lemma 6, there exists a system of local quasi-coordinates $\mathcal{V} = \{(V_\alpha; v_\alpha^1, v_\alpha^2)\}_\alpha$ of $X'' \setminus D''$ and a neighborhood U

of Supp(D'') such that, for all $x \in D''$, there is a local uniformizing (orbifold) chart (B, G_x, π_x), with $U_x = \pi_x(B) \subset U$. Moreover, we have

a) $(\cup_\alpha(\text{Image}(V_\alpha))) \cup U = X''$,
b) $(\cup_\alpha(\overline{\text{Image}(V_\alpha)})) \cap \text{Supp}(D'') = \phi$,
c) if the image of V_α does not intersect U, then the quasi-coordinates (v_α^1, v_α^2) are local coordinates in the usual sense,
d) there are positive numbers ε, δ, independent of V_α in $\mathcal{V}$ such that $V_\alpha \subset \mathbb{C}^2$ contains a ball of radius ε and is contained in a ball of radius δ,
e) on a quasi-coordinate chart $(V_\alpha; v_\alpha^1, v_\alpha^2)$, there are positive constants c and $\mathcal{A}_k$ $(k = 0, 1, 2 \ldots)$ such that we have the inequality between Hermitian matrices

$$c^{-1}(\delta_{ij}) < (g_{\alpha i \bar{j}}) < c(\delta_{ij}),$$

and

$$|\partial^{|p|+|q|} g_{\alpha i \bar{j}} / \partial v_\alpha^p \partial \bar{v}_\alpha^q| < \mathcal{A}_{|p|+|q|}$$

for all multi-indices p, q, where

$$\mu_*(\omega) = \sqrt{-1} \sum_{i,j} g_{\alpha i \bar{j}} dv_\alpha^i \wedge d\bar{v}_\alpha^j$$

in terms of (v_α^1, v_α^2), the same type of inequalities also holding for the metrics on B lifted from the U_x, $x \in D''$: see [82], pp. 393–394.

The last condition is known as "bounded geometry", so a quasi-coordinate system for (X'', D'', ω) exhibits (X'', D'') as a complete Kähler orbifold with bounded geometry. The approximations of our function g (not to be confused with the notation $g_{\alpha i \bar{j}}$ for the coefficients of a metric) at the ends of X_0'' are C^∞ in the following sense: the metric $\mu_*(\omega)$, dependent on g, and the metric induced by the invariant metric in the local uniformization defined at an end are of bounded geometry at the same time: both metrics satisfy (e) with respect to the same constants and quasi-coordinate system.

We now proceed in a manner similar to that of the discussion of the existence of a Kähler-Einstein metric in Chapter 4, but adapted to the orbifold case. Instead of introducing a function f with $\Psi = e^f \omega^2$ that is smooth, we need to place different "smoothness" conditions on f that apply to "b-functions." The relevant notion, developed in [81], [84], is that of the Banach space of $b - C^{k,\alpha}$ functions on (X'', D'', ω), where k is a nonnegative integer and α is in the open interval $(0, 1)$. (This α should not be confused with the index α used in the definition of quasi-coordinates.) This is quite technical, and for precise details we refer the reader to the references just cited. They use the existence of a suitable quasi-coordinate system to which a function can be lifted on $X'' \setminus D''$, and the coordinates of the local uniformizations to which a function can be lifted on D''. The $C^{k,\alpha}$ refers to a Hölder condition whereby functions have continuous partial derivatives D^k up to order k and the kth partial derivatives are Hölder continuous with exponent α, where $0 < \alpha < 1$, that is, $|D^k f(z) - D^k f(z')| \leq C|z - z'|^\alpha$ for z, z' in a suitable domain and some constant $C > 0$. In other words, the integer k refers to the order of the partial derivatives of the lifted function that exist and are bounded, and the α refers to bounding the difference of these derivatives, at points with respective local coordinates z and z', multiplied by $|z - z'|^{-\alpha}$. One shows that $\Psi = e^f \omega^2$ where $f \in b - C^{k,\alpha}$, for all integers $k \geq 0$ and all $\alpha \in (0, 1)$. Using arguments analogous to this of Chapter 4, the existence of a Kähler-Einstein b-metric (i.e., orbifold metric) $\check{\omega} = \omega + \partial\bar{\partial}u$ reduces to the Monge-Ampère equation, with $\lambda = -1$ (here given in terms of $(1, 1)$-forms and not the coefficients $g_{i\bar{j}}$ as in Chapter 4),

$$\left(\omega + \partial\bar{\partial}u\right)^2 = e^{u+f}\omega^2.$$

Using [84], p. 357, and the references cited there, it can be shown that there is a solution $u \in \cap_{k,\alpha}(b - C^{k,\alpha})$ such that $\check{\omega}$ induces a Kähler-Einstein b-metric on (X_0'', D_0'') that satisfies certain a priori estimates ensuring that there is a positive constant c with

$$c\omega < \check{\omega} < c^{-1}\omega.$$

This implies that the singular differential form $\check{\omega}$, which by construction lives on X, induces a complete Kähler-Einstein orbifold metric on (X_0'', D_0'') with negative scalar curvature.

To derive the orbifold version of the Miyaoka-Yau inequality, we now use the Chern forms $\check{\gamma}_1$, $\check{\gamma}_2$ for the metric $\check{\omega}$. Using the same arguments as in Chapter 4, §4.3, and the fact that $\check{\omega}$ is a Kähler-Einstein b-metric, we have a pointwise inequality of $b-$ 2-forms $0 \leq 3\check{\gamma}_2 - \check{\gamma}_1^2$ that, when integrated over the normal compact surface X'' in a way that allows for the b-space structure on (X_0'', D_0''), gives the appropriate Miyaoka-Yau inequality. Moreover, as in Chapter 4, the difference $3\check{\gamma}_2 - \check{\gamma}_1^2$ is the square of the pointwise deviation of the Kähler-Einstein b-metric $\check{\omega}$ from a constant holomorphic sectional curvature. Therefore, if we have equality in the Miyaoka-Yau inequality, adapting the arguments of [81], pp. 67–69, we see that the geodesic developing map associated to $\check{\omega}$ exhibits X'' as a compactified ball quotient $\overline{\Gamma \setminus B_2}$ with $X_0'' \simeq \Gamma \setminus B_2$. Moreover, the natural map $B_2 \to X_0''$ has branching of order $n_i > 0$ along $D_{i,0}''$ and $m_j > 0$ along $E_{j,0}''$, as required.

We now give some more details, closely based on the papers [80], [81], [82], [84] of R. Kobayashi. Indeed, following [84], it suffices to adapt the treatment in [82] of Kähler-Einstein metrics and the Miyaoka-Yau inequality for open Satake V-surfaces, that is, open complex surfaces with a finite number of isolated quotient singularities, to the case of the orbifold (X_0'', D_0''), whose singularities are not isolated, but lie along the irreducible components of D_0''. Roughly speaking, this amounts to allowing for the effect of the weight data in (X, D) on the Chern classes of (X_0'', D_0'').

We have constructed a $b - (1, 1)$-form $\mu_*(\omega)\,|_{X_0''}$ on (X_0'', D_0''), where ω lives on (X, D), such that, for each admissible (k, α), the $(1, 1)$-form ω is a $C^{k,\alpha}$-form on

$$(X, D)_{\text{reg}} := X \setminus \text{Supp}(D) \simeq X'' \setminus \text{Supp}(D'')$$

$$\simeq X_0'' \setminus \text{Supp}(D_0'') =: (X_0'', D_0'')_{\text{reg}},$$

where the subscript "reg" denotes the regular points of the corresponding orbifold. We have $\mathring{\omega} = \omega + \partial \bar{\partial} u$, so we can work with ω as integrating the Chern forms of $\mathring{\omega}$ over (X_0'', D_0'') will give the same result as integrating those of ω by [81], Lemma 9; see also [84], p. 357. On $(X_0'', D_0'')_{\text{reg}}$, we denote $\mu_*(\omega)$ by ω. Let x be a point in $(X_0'', D_0'')_{\text{reg}}$. In a neighborhood $U_x \subseteq (X_0'', D_0'')_{\text{reg}}$ of x, let (g_{ij}) be the Hermitian matrix associated to ω. We define the Chern forms of ω on this open set in the usual way. Namely, let ν be the matrix of 1-forms,

$$\nu = (\partial g)g^{-1},$$

and let Θ be the curvature matrix consisting of the matrix of $(1, 1)$-forms,

$$\Theta = \bar{\partial}\nu.$$

Then, as in Chapter 4, we have the Chern forms defined on this open set by

$$\gamma_1 = (\sqrt{-1}/2\pi)(\Theta_1^1 + \Theta_2^2) = (\sqrt{-1}/2\pi)\text{Trace}(\Theta),$$

$$\gamma_2 = (-1/8\pi^2)(\Theta_1^1 \wedge \Theta_2^2 - \Theta_2^1 \wedge \Theta_1^2) = (-1/8\pi^2)\text{Det}(\Theta).$$

Consider an orbifold chart (B, G, π) on (X_0'', D_0''), where G is a finite subgroup of the automorphisms of the complex ball B. This includes the case where G is trivial. As $\mu_*(\omega)|_{X_0''}$ is a $b - (1, 1)$-form on (X_0'', D_0''), the form $\pi^*(\omega|_{\pi(B) \cap (X_0'', D_0'')_{\text{reg}}})$ extends to a nonsingular $(1, 1)$-form $\widehat{\omega}$ on B, and the associated Chern forms $\widehat{\gamma}_1, \widehat{\gamma}_2$ are extensions to B of the pullback by π of γ_1, γ_2 on $\pi(B) \cap (X_0'', D_0'')_{\text{reg}}$. Define the orbifold integral of a $b - 2$-form supported on a compact subset of $\pi(B)$ by

$$\int_{\pi(B)}^{\text{orb}} \eta := \frac{1}{|G|} \int_B \pi^*(\eta).$$

Recall that X' denotes the compact complex surface, with, at worst, isolated finite quotient singularities, obtained from X by blowing down the D_t and E_s with negative weights. Let D'_∞ denote the image in X' of the union of the D_t and E_s with infinite weight. By assumption, the irreducible components of D'_∞ are disjoint and lie in the set of regular points of X'. Let $D'_{+,0}$ denote the intersection of $X' \setminus D'_\infty$ with the image in X' of

$$\sum_{n_i > 0} \left(1 - \frac{1}{n_i}\right) D_i + \sum_{m_j > 0} \left(1 - \frac{1}{m_j}\right) E_j.$$

Our X' corresponds to the X of [82]. Therefore, $X' \setminus D'_\infty \simeq X''_0$ is a Satake V-surface in the sense of [82], which is open when D'_∞ (denoted by C in [82]) is nonempty. The minimal resolution of X' is X (denoted by $\overline{X}$ in [82]), whose corresponding exceptional divisor D_- is the union of all the D_t and E_s of negative weight (our D_- is denoted by D, with irreducible components D_i, in [82]). We have an isomorphism of orbifolds

$$(X' \setminus D'_\infty, D'_{+,0}) \simeq (X''_0, D''_0),$$

so that we have the same situation as in [82], Theorems 1 and 2, except that we have to allow for the weight data along all the D_i and E_j. Note that the

$$K_{\overline{X}} + \sum_i \mu_i D_i + C$$

of [82] corresponds to our $K_{X''} + D''$. Despite these differences, we may take advantage of the computations of [81], Lemma 9, and [82], Lemma 12, to deal with the integration of the Chern forms near a singularity of the support of $D'_{+,0}$, and near the boundary of $X' \setminus D'_\infty$. As ω is approximately invariant near the ends of X''_0, we can use a method similar to that of [52], described in Chapter 4, §4.3, to compute the contribution of the cusps. The difference between ω and the invariant metric is explained in somewhat more detail in our earlier discussion of quasi-coordinate systems and bounded geometry. For full details, again see [81], [82], [84]. The definition of orbifold integration for a chart (B, G, π), with G finite, is all that is required to deal with the fact that the singularities of the orbifold $(X' \setminus D'_\infty, D'_{+,0})$ are not isolated.

As remarked at the bottom of [84], p. 357, it follows from the construction of ω and the *a priori* estimates for u that the singular differential form $\check{\omega}$ defines a real closed $(1, 1)$-current on X whose cohomology class is

$$c_1(\mu^*(K_{X''} + D'')) = (K_{X''} + D'')^2 := (K_X + D)^2.$$

The proof that this is the orbifold integral over (X_0'', D_0'') of the first Chern form of ω is essentially as in [82], Lemma 12, which in turn uses [81], Lemma 9.

We now discuss the orbifold integral of the second Chern form of the b-metric ω by adapting the arguments of [82], §4, following [84], §3.2.3, specialized to our situation. An orbifold chart (B, π, G_P) on (X_0'', D_0''), where G_P is finite and fixes the origin $0 \in B$, and where $P = G_P \cdot 0 \in X_0''$, plays the same role as the local uniformization of an isolated finite quotient singularity in [82]. In [82], the "V-metric" lifts to a nonsingular metric on the local uniformizations, just as our b-metric ω lifts to a nonsingular metric on the local orbifold charts. This is why, formally, the situations are very similar. We also need to deal with the ends of X_0'', but, again, all the requisite tools are in [82].

The contribution to the orbifold integration of the second Chern form of ω coming from the $D_{i,0}''$ (weight $n_i > 0$) and $E_{j,0}''$ (weight $m_j > 0$), minus their singular intersection points, can be computed using, essentially, the same arguments as in Chapter 3, §3.1, applied to the local orbifold charts. At the regular points of the divisors, the covering group of the orbifold chart is a finite cyclic group of order the weight associated to the divisor, and at a double intersection point it is the product of two such groups. This gives the following contribution to the overall integral of the second Chern form,

$$\sum_{n_i > 0} \frac{1}{n_i} \left(e(D_{0,i}'') - d_i \right) + \sum_{m_j > 0} \frac{1}{m_j} \left(e(E_{0,j}'') - e_j \right) + \sum_P \frac{1}{|G_p|}.$$

Here, d_i is the number of intersection points of the support of D_0'' on D_i, and e_j is the number of intersection points of the support of D_0'' on E_j. The point P runs over the nonsingular (double) intersection points of the support of D_0''. If $P = D_{i,0}'' \cap D_{\ell,0}''$, then $|G_P| = n_i n_\ell$, and if $P = D_{i,0}'' \cap E_{j,0}''$, then $|G_p| = n_i m_j$.

At a singular intersection point P of the support of D_0'', we can essentially apply the same arguments as in Chapter 4, §4.2, and [82], §4. Consider a local orbifold chart (B, G_P, π), with $P = G_P \cdot 0$ and a resolution $(\overline{U}, E_P)$ of (U, P), where $U = \pi(B)$. The contribution to the integral of the second Chern class of ω is

$$e(\overline{U}) - \left(e(E_P) - \frac{1}{|G_P|} \right) = e(\overline{U} \setminus E_P) + \frac{1}{|G_P|} = e(U \setminus \{P\}) + \frac{1}{|G_P|}.$$

Let P be a point of $X'' \setminus X_0''$. Then P is an elliptic cusp singularity and is resolved by an elliptic curve C_P. We can argue in a similar fashion as in Chapter 4, §4.3, since, by construction, the metric associated to ω is approximately the invariant ball metric. For a suitable neighborhood U of P, let $(\overline{U}, C)$ be a resolution of P. We have a contribution to the integral of the second Chern class given by

$$e(\overline{U}) - e(C) = e(\overline{U} \setminus C) = e(U \setminus \{p\}) = e(U) - 1.$$

Let S be the set of singular points of (X_0'', D_0''). Putting all the above remarks together, we see that the integral of the second Chern class of ω gives

$$e(X_0'' \setminus S) + \sum_{n_i > 0} \frac{1}{n_i} \left(e(D_{0,i}'') - d_i \right) + \sum_{m_j > 0} \frac{1}{m_j} \left(e(E_{0,j}'') - e_j \right) + \sum_P \frac{1}{|G_P|},$$

where now P ranges over all the log-*terminal* singularities of (X'', D''), that is, over the intersection points of the support of D_0'' (in our case), and G_P is the local fundamental group at P in the sense of orbifolds or b-spaces as described above. We can rewrite this as

$$e(X_0'') + \sum_{n_i > 0} \left(\frac{1}{n_i} - 1 \right) \left(e(D_{0,i}'') - d_i \right)$$

$$+ \sum_{m_j > 0} \left(\frac{1}{m_j} - 1 \right) \left(e(E_{0,j}'') - e_j \right) + \sum_P \left(\frac{1}{|G_P|} - 1 \right),$$

as in [84], §3.2.3 (applied to our situation). (If we did not take into account the contribution of the $D_{i,0}''$ and $E_{j,0}''$ of positive weight, then the above formulas for the integral of the first and second Chern classes agree with those of [82], Lemma 12.)

From the above discussion, we see that the Miyaoka-Yau inequality is

$$\frac{1}{3} \left(K_{X''} + D'' \right)^2 \leq e(X_0'') + \sum_{n_i > 0} \left(\frac{1}{n_i} - 1 \right) \left(e(D_{0,i}'') - d_i \right) +$$

$$+ \sum_{m_j > 0} \left(\frac{1}{m_j} - 1 \right) \left(e(E_{0,j}'') - e_j \right) + \sum_P \left(\frac{1}{|G_P|} - 1 \right),$$

where P runs over the intersection points of D_0'', and where d_i is the number of intersection points of D'' above $D_{0,i}''$, and e_j is the number of intersection points of D'' above $E_{0,j}''$.

When we have equality in the above inequality, the orbifold (X_0'', D_0'') is isomorphic to (B_2, Γ, π), where $\Gamma \backslash B_2 \simeq X_0''$ and π is the natural map from B_2 to $\Gamma \backslash B_2$. By this we mean that π gives an infinite ramified cover $B_2 \to X_0''$ such that the b-space structure of (X_0'', D_0'') is given by b_π. The point compactification (adding one point at each cusp of Γ) is denoted by $\overline{\Gamma \backslash B_2}$ and is isomorphic to X''. Each of these cusps is resolved by an elliptic curve.

Using a theorem of Borel and Selberg [18], [129], we deduce that the group Γ has a torsion-free normal subgroup Γ' of finite index such that the smooth surface $\Gamma' \backslash B_2$ is isomorphic to a surface described by a $Y' \backslash C$ as above (see Chapter 5, §5.6, and the remarks at the beginning of this section). To see why elliptic curves resolve the cusps of Γ, see [52], [145], and also the following discussion of isotropy groups, as well as Chapter 5, §5.4. □

6.8 ISOTROPY SUBGROUPS OF THE COVERING GROUP

Consider the isotropy groups of $\Gamma' \backslash \Gamma$ acting on $\Gamma' \backslash B_2 \simeq Y' \backslash C$, where Γ' is a normal torsion-free subgroup of Γ. Clearly, $\Gamma' \backslash B_2$ is a covering of $\Gamma \backslash B_2$, of degree $N = |\Gamma' \backslash \Gamma|$. By a result of Chevalley [28], a point x is a smooth point of $\Gamma'' \backslash B_2$, where Γ'' is a discrete subgroup of the automorphisms of the ball, if and only if the isotropy group Γ_z'', where z is any lift to B_2 of x, is a (complex) reflection group. Recall that a (complex) reflection is an element of $GL(n, \mathbb{C})$ with one eigenvalue equal to a nontrivial root of unity and the other eigenvalues equal to 1. A reflection group is one generated by reflections. As Γ'' acts properly discontinuously, the isotropy group Γ_z'' of any point z in B_2 is isomorphic to a finite subgroup of $GL(2, \mathbb{C})$. Therefore, the isotropy groups of $\Gamma' \backslash \Gamma$ acting on $\Gamma' \backslash B_2$, all of whose points are smooth, are conjugate to finite subgroups of $U(2)$ generated by complex reflections.

We now list the possibilities for the isotropy groups of Γ acting on B_2 by looking at their relation to the b-space structure of (X_0'', D_0''). Our main reference is [145], Chapter 11, to which we refer the reader for full details. Let π be the natural map $\pi : B_2 \to \Gamma \backslash B_2 \simeq X_0''$. There is an associated "b-map," in the sense introduced earlier in this section, given by $b_\pi : x \in X_0'' \to |\Gamma_z|$, for any $z \in \pi^{-1}(x)$ (see [75], [138]). This is clearly independent of our choice of $z \in \pi^{-1}(x)$. The pair (X_0'', b_π) is the corresponding b-space.

The irreducible finite reflection groups have been classified by Shepherd and Todd [132]. In our situation, we can readily identify which finite groups G occur in the local finite orbifold uniformizing charts $(\mathbb{C}^2, G)$, where each element of G fixes the origin $(0, 0) \in \mathbb{C}^2$. Let $x \in X_0''$ and $(\mathbb{C}^2, G)$ be a local orbifold uniformizing chart centered at $x = G \cdot (0, 0)$. The group G is nontrivial only for points x in the support $\mathrm{Supp}(D_0'')$ of D_0''. The image PG of G in the projective group $PGL(2, \mathbb{C})$ gives rise to a central extension $P : G \to PG$ with a kernel generated by multiples of the identity matrix I_2.

As G is finite, the kernel of P is cyclic and generated by ζI_2, where ζ is a root of unity. As $x = G \cdot (0, 0)$, the group PG is isomorphic to Γ_z, for any $z \in \pi^{-1}(x)$, and the different choices of z give rise to conjugate Γ_z.

Let x be a point on an irreducible component of $\text{Supp}(D_0'')$ of weight $b > 0$, which is not an intersection point. The group G is isomorphic to the finite cyclic group $\mathbb{Z}/b\mathbb{Z}$, which is an abelian reflection group. For a simple intersection point of an irreducible component of weight $b_1 > 0$ and one of weight $b_2 > 0$, the group G is the abelian reflection group $(\mathbb{Z}/b_1\mathbb{Z}) \times (\mathbb{Z}/b_2\mathbb{Z})$, as in our definition of "good covering" in Chapters 3 and 5.

A point on $\text{Supp}(D_0'')$ that is the image of the blow-down of a divisor on X given by E_s, with weight $m_s < 0$, or D_t, with weight $n_t < 0$ and $\sigma_t = 2$, is a smooth point of X_0'' at the intersection of three irreducible components of $\text{Supp}(D_0'')$ of respective weights $p, q, r > 0$. By the diophantine conditions (6.3) of this chapter,

$$(1/p + 1/q + 1/r) - 1 =: 1/\rho > 0,$$

where $2\rho = |m_s|$ (for E_s) or $|n_t|$ (for D_t). The group G is the maximal reflection group $\langle p, q, r \rangle_\rho$ of order $4\rho^2$; see [145], Chapter 11. It is a central extension of order 2ρ of the spherical triangle group of signature (p, q, r), which is itself of order 2ρ (for a discussion of spherical triangle groups, see Chapter 2). When $n_t < 0$ and $\sigma_t = 3$, we have a singular point of X_0'', so that G is no longer a reflection group. Again, from [145], Chapter 11, the group G is a cyclic extension of the maximal reflection group $\langle p, q, r \rangle_\rho$ of order $4\rho^2$. We have $4\rho^2 = (|n_t|/2)^2$, again by the diophantine conditions (6.3) of this chapter. Referring to §4.2, we see that the contraction of a -2 curve gives rise to a cyclic quotient singularity of order 2. Therefore G has order twice that of $\langle p, q, r \rangle_\rho$. The action of the nontrivial element of the cyclic group $\mathbb{Z}/2\mathbb{Z}$ on $\mathbb{C}^2$ is given by $(x, y) \mapsto (-x, -y)$. The orbit space of $(\mathbb{C}^2, \mathbb{Z}/2\mathbb{Z})$ is the hypersurface of $\mathbb{C}^3$ given by $z^2 = xy$, which has a cyclic quotient singularity at the origin (see [138]). Notice that, by the discussion at the beginning of Chapter 5, §5.6, we see again that the order of G is m_s^2 or $n_t^2/(\sigma_t - 1) = (n_t/(\sigma_t - 1))^2(\sigma_t - 1)$.

If there are divisors of infinite weight on X, then X_0'' is not compact, and we have the point compactification $X'' = \overline{\Gamma \backslash B_2}$, which may have singularities. In fact, the singularities occur precisely at the images of the contracted D_t with negative or infinite weight and $\sigma_t > 2$. When $n_t = \infty$, by the discussion of Chapter 5, §5.3, we either have $\tau_t = 3$, with three irreducible components of $\text{Supp}(D_0'')$ meeting at the contracted D_t having weights $(2, 4, 4)$, $(3, 3, 3)$, or $(2, 3, 6)$, or we have $\tau_t = 4$, with four components meeting at the contracted point each having weight 2. This occurs for the configuration Ceva(3) with $n = \infty$ discussed in Chapter 5, §5.6, and corresponds to entry number 23 of Table 5.12; see also [145], Chapter 11, p. 183. The image p on X'' of the contraction of an infinite weight divisor corresponds to a point z of the boundary of B_2 with isotropy group $G = \Gamma_z$ a discrete subgroup of a parabolic group. The group G is a central extension of a 1-dimensional crystallographic group. The crystallographic group is itself a central extension of a lattice in $\mathbb{C}$ by a finite cyclic group leaving the lattice invariant. This finite cyclic group is called the point group. The image p of a contracted divisor of infinite weight is smooth if and only if $G = \Gamma_z$, where z is a point on the boundary of B_2 corresponding to p, is a reflection group (see [145], p. 174). In that case, the group G is a maximal reflection group with projectivization the crystallographic groups given by the elliptic triangle groups of signature $(2, 4, 4), (3, 3, 3), (2, 3, 6)$ or the group $\langle 2, 2, 2, 2; w \rangle$, which has presentation

$$M_1^2 = M_2^2 = M_3^2 = M_4^2 = M_1 M_2 M_3 M_4 = 1$$

and parameter w in the complex upper half plane. For this last group, the action of the point group on the lattice corresponds to the automorphism -1 of $\mathbb{Z} + \mathbb{Z}w$. In all cases, the quotient of $\mathbb{C}$ by the crystallographic group is a rational curve (corresponding to the rational curve of infinite weight that is contracted). If H is one of these crystallographic groups, the corresponding maximal reflection group G is denoted by $\langle p, q, r \rangle_\infty$, $1/p + 1/q + 1/r = 1$, with presentation

$$M_1^p = M_2^q = M_3^r = [M_1 M_2 M_3, M_j] = 1, \qquad j = 1, 2, 3,$$

or $\langle 2, 2, 2, 2; w \rangle_\infty$, with presentation

$$M_j^2 = [M_1 M_2 M_3 M_4, M_j] = 1, \qquad j = 1, 2, 3, 4.$$

When $\sigma_t > 2$, we contract a curve D_t of self-intersection $1 - \sigma_t$, so that (again comparing to §4.2, or using §5.6), the group G is a central extension of a maximal reflection group by a cyclic group of order $\sigma_t - 1$. The curve D_t is the smooth compactification of the singularity resulting from its contraction. If Γ' is a normal subgroup of finite index in Γ acting without fixed points in B_2, the isotropy group of Γ' at a boundary point z of B_2, corresponding to the image p of a contracted curve of infinite weight, has a trivial point group, and therefore its corresponding crystallographic group is a lattice L. In that case, there is a smooth completion of $\Gamma'_z \backslash B_2$ obtained by adding the elliptic curve $\mathbb{C}/L$ (see [145], p. 173). Therefore $\Gamma' \backslash B_2$, compactified by these elliptic curves is Y'. The quotient of Y' by $\Gamma' \backslash \Gamma$ is X'. If, as before, $\overline{\Gamma \backslash B_2}$ and $\overline{\Gamma' \backslash B_2}$ denote the compactifications by single points at the cusps, then $\overline{\Gamma' \backslash B_2}$ is Y', with the elliptic curves in C collapsed to points, and $\overline{\Gamma \backslash B_2}$ is X''.

Chapter Seven

Appell Hypergeometric Functions

In a short note, written in 1880 [4], and in a memoir, written in 1882 [5], Appell introduced four series F_1, F_2, F_3, F_4 in two complex variables, each of which generalizes the classical Gauss hypergeometric series and satisfies its own system of two linear second order partial differential equations. The solution spaces of the systems corresponding to the series F_2, F_3, F_4 all have dimension 4, whereas that of the system corresponding to the series F_1 has dimension 3. The extension of Appell's results to the n-variable case, $n \geq 3$, was done by Lauricella [93]. The work of Appell and Lauricella is included in the extensive treatment of hypergeometric functions in several variables published in 1926 by Appell and Kampé de Fériet [6].

Work of Goursat [47] and Picard [116] in the early 1880s complemented the work of Appell and showed that the two-variable analogue of the classical Riemann problem leads naturally to the function F_1. The singularities of the system of partial differential equations satisfied by $F_1 = F_1(x, y)$ lie along the seven lines $x, y = 0, 1, \infty$ and $x = y$. Picard characterized the solutions of the F_1-system as the multivalued functions of two variables with exactly three linearly independent branches and with prescribed ramification along these seven lines, the series F_1 being a solution that is holomorphic and takes the value 1 at $(x, y) = (0, 0)$. Compare this with Riemann's characterization of the Gauss hypergeometric function in terms of the branching behavior at the singular points $0, 1, \infty$ referred to in Chapter 2.

In this chapter, we restrict our attention to the F_1-system. Under certain conditions, its monodromy group acts on the complex 2-ball. Building on earlier work of Picard [116] and Terada [136], Deligne and Mostow [33] determined sufficient conditions for this action to be discontinuous. They also determined which of these discontinuously acting groups is arithmetic. The remaining ones provide rare examples of non-arithmetic groups acting discontinuously on an irreducible bounded complex symmetric domain of dimension greater than 1.

The space of regular points for the system of partial differential equations for F_1 is biholomorphically equivalent to an open subset of the algebraic surface X obtained by blowing up the four triple points of the complete quadrilateral arrangement in $\mathbb{P}_2(\mathbb{C})$. The singular hypersurfaces of the F_1-system correspond to ten divisors on X coming either from the lines of the complete quadrilateral or from the blown-up points. In Chapters 5 and 6, we constructed good coverings Y of X ramified along these ten divisors and satisfying $\text{Prop}(Y) = 0$. We also considered a generalization of this situation by allowing weights along the ten divisors on X that could be negative or infinite. The resulting surfaces are isomorphic to quotients of the complex 2-ball by subgroups of finite index in the monodromy group of an F_1-system. They are two-dimensional generalizations of finite coverings of $\mathbb{P}_1$, branched over 0, 1, ∞, given by quotients of the upper half plane by subgroups of finite index in monodromy groups of Gauss hypergeometric functions.

7.1 THE ACTION OF S_5 ON THE BLOWN-UP PROJECTIVE PLANE

In Chapters 3 and 5 we studied the algebraic surface X obtained by blowing up the four triple points of the complete quadrilateral arrangement in $\mathbb{P}_2$. The surface X has ten exceptional curves $D_{\alpha\beta} = D_{\beta\alpha}$, where $\alpha \neq \beta$ and $\alpha, \beta \in \{0, 1, 2, 3, 4\}$. The symmetric group S_5 operates on the set of curves $D_{\alpha\beta}$, preserving the intersection numbers of each pair. However, a priori there is no reason why S_5 should operate on all of X. That this is the case will become apparent in light of an alternative description of X given in this section. In Figure 1 of Chapter 3, the four curves obtained by blowing up the triple points of the complete quadrilateral are

$$D_{01}, \quad D_{02}, \quad D_{03}, \quad D_{04}.$$

For any α, the curves $D_{\alpha\beta}, D_{\alpha\gamma}, D_{\alpha\delta}, D_{\alpha\varepsilon}$ are disjoint if $(\alpha, \beta, \gamma, \delta, \varepsilon)$ is a permutation of $(0, 1, 2, 3, 4)$. Thus, we have five possible ways of blowing down four disjoint exceptional curves on X to obtain $\mathbb{P}_2$. These correspond to the five possible values of α. As we saw in Chapter 4, §4.1, $\mathbb{P}_1 \times \mathbb{P}_1$ with one point blown up is isomorphic to $\mathbb{P}_2$ with two points blown up. Therefore, it must be possible to blow down three disjoint exceptional curves on X to recover $\mathbb{P}_1 \times \mathbb{P}_1$. In fact, for a permutation $(\alpha, \beta, \gamma, \delta, \varepsilon)$ of $(0, 1, 2, 3, 4)$, the curves $D_{\gamma\delta}, D_{\delta\varepsilon}, D_{\gamma\varepsilon}$ are disjoint exceptional curves. They are the three exceptional curves intersecting $D_{\alpha\beta}$ in X. Blowing down these three curves yields $\mathbb{P}_1 \times \mathbb{P}_1$. Thus, there are ten possible ways to blow down three curves in X and obtain $\mathbb{P}_1 \times \mathbb{P}_1$. We shall now do this explicitly. Indeed, $D_{\alpha\beta} = D_{\beta\alpha}$ belongs to both of the following sets of four disjoint curves on X:

$$D_{\alpha\beta}, \quad D_{\alpha\gamma}, \quad D_{\alpha\delta}, \quad D_{\alpha\varepsilon}$$

and

$$D_{\beta\alpha}, \quad D_{\beta\gamma}, \quad D_{\beta\delta}, \quad D_{\beta\varepsilon}.$$

Blowing down $D_{\alpha\beta}$ on X gives a diagram of nine curves.

Blowing down either

$$D_{\alpha\gamma}, \quad D_{\alpha\delta}, \quad D_{\alpha\varepsilon}$$

or

$$D_{\beta\gamma}, \quad D_{\beta\delta}, \quad D_{\beta\varepsilon}$$

in the resulting hexagon gives $\mathbb{P}_2$ with the four triple points of a complete quadrilateral. We denote these two copies of the projective plane by A and B. They are related by a classical Cremona transformation as follows: blow up the vertices formed by a "triangle" of three lines in general position and then blow down the resulting hexagon at the proper transforms of the three lines of the original triangle. The lines in A passing through the point resulting from the blow-down of $D_{\alpha\beta}$ define a fibration of X by rational curves. The fibers are parameterized by $D_{\alpha\beta}$, which is itself isomorphic to $\mathbb{P}_1$. The fibers are smooth, except for the three exceptional fibers, given by

$$D_{\gamma\delta} \cup D_{\alpha\varepsilon}, \; D_{\delta\varepsilon} \cup D_{\alpha\gamma}, \; D_{\gamma\varepsilon} \cup D_{\alpha\delta}.$$

The mapping

$$\pi_\alpha : X \to \mathbb{P}_1$$

defining this fibration can be normalized by requiring that the three exceptional fibers, in the above order, be mapped to the points $0, 1, \infty$. In the same way, using the projective plane B, we can define a mapping

$$\pi_\beta : X \to \mathbb{P}_1,$$

normalized so that

$$D_{\gamma\delta} \cup D_{\beta\varepsilon}, \; D_{\delta\varepsilon} \cup D_{\beta\gamma}, \; D_{\gamma\varepsilon} \cup D_{\beta\delta}$$

are mapped, in this order, to $0, 1, \infty$. The maps π_α and π_β depend on the permutation $(\alpha, \beta, \gamma, \delta, \varepsilon)$ of $(0, 1, 2, 3, 4)$ that was fixed at the outset. The conics through the points given by the blow-down of $D_{\alpha\beta}, D_{\alpha\gamma}, D_{\alpha\delta}, D_{\alpha\varepsilon}$ in the projective plane A correspond to the lines through the point given by the blow-down of $D_{\beta\alpha}$ in the projective plane B. The map

$$\pi = \pi_\alpha \times \pi_\beta : X \to \mathbb{P}_1 \times \mathbb{P}_1$$

is a blow-down of X to $\mathbb{P}_1 \times \mathbb{P}_1$, collapsing $D_{\gamma\delta}, D_{\delta\varepsilon}, D_{\gamma\varepsilon}$ to $(0, 0)$, $(1, 1)$, (∞, ∞), where $D_{\alpha\varepsilon}, D_{\alpha\gamma}, D_{\alpha\delta}$ become $\mathbb{P}_1 \times \{0\}$, $\mathbb{P}_1 \times \{1\}$, $\mathbb{P}_1 \times \{\infty\}$ and $D_{\beta\varepsilon}, D_{\beta\gamma}, D_{\beta\delta}$ become $\{0\} \times \mathbb{P}_1$, $\{1\} \times \mathbb{P}_1$, $\{\infty\} \times \mathbb{P}_1$. Clearly, the image of $D_{\alpha\beta}$ under π is the diagonal of $\mathbb{P}_1 \times \mathbb{P}_1$. The lines in A passing through $D_{\alpha\beta}$ correspond to the curves $\{t\} \times \mathbb{P}_1$ in $\mathbb{P}_1 \times \mathbb{P}_1$. The lines in B passing through $D_{\beta\alpha}$ correspond to the curves $\mathbb{P}_1 \times \{t\}$.

In the sequel, we shall use an alternate description of the space X.

Definition 7.1 *Define X' to be the smooth two-dimensional complex manifold given by*

$$X' = \{(z_0, z_1, z_2, z_3, z_4) \in \mathbb{P}_1^5 : \text{no three components are equal}\}/\text{Aut}(\mathbb{P}_1),$$

where $\text{Aut}(\mathbb{P}_1)$ *acts diagonally and freely.*

There is a natural action of S_5 on $\mathbb{P}_1^5$ that descends to X'. We now show how X' can be identified with X, so the S_5-action on X becomes apparent a fortiori. In order to give the identification of X with X' explicitly, it is convenient to use the cross ratio with values in $\mathbb{P}_1$,

$$V(z) = V(z, z_\gamma, z_\delta, z_\varepsilon) = \frac{(z - z_\varepsilon)(z_\gamma - z_\delta)}{(z - z_\delta)(z_\gamma - z_\varepsilon)}. \tag{7.1}$$

The latter is defined for $z, z_\gamma, z_\delta, z_\varepsilon \in \mathbb{P}_1$, when no three of these points are equal. The cross ratio is invariant under the diagonal action of $\text{Aut}(\mathbb{P}_1)$. There are other conventions for the cross ratio; however, using them does not affect the discussion that follows. On X' there are ten curves $D'_{\alpha\beta}$ given by the image on X' of $z_\alpha = z_\beta$ for $\alpha \neq \beta$ and $\alpha, \beta \in \{0, 1, 2, 3, 4\}$. Just as for the $D_{\alpha\beta}$, we have the intersection behavior

$$D'_{\alpha\beta} \cap D'_{\gamma\delta} = \begin{cases} \phi & \{\alpha\beta\} \cap \{\gamma\delta\} \neq \phi \\ \{\text{a point}\} & \{\alpha\beta\} \cap \{\gamma\delta\} = \phi. \end{cases}$$

These curves are isomorphic to $\mathbb{P}_1$: consider (7.1) applied to $z = z_\alpha = z_\beta$,

$$\{z_\alpha = z_\beta\} \mapsto V(z_\alpha). \tag{7.2}$$

This induces a map of the curve $D'_{\alpha\beta}$ on X' to $\mathbb{P}_1$. For each fixed set of pairwise distinct points, $z_\gamma, z_\delta, z_\varepsilon$, the map $z \mapsto V(z)$ given by (7.1) is a fractional linear transformation of z with nonzero determinant

$$(z_\gamma - z_\delta)(z_\gamma - z_\varepsilon)(z_\varepsilon - z_\delta).$$

Consider the open subset of $D'_{\alpha\beta}$ that excludes the intersection points with the $D'_{\gamma\delta}$, $\{\alpha\beta\} \cap \{\gamma\delta\} = \phi$. Its image under (7.2) is $\mathbb{P}_1 \setminus \{0, 1, \infty\}$. On the other hand, the intersection points

$$D'_{\alpha\beta} \cap D'_{\gamma\delta}, \ D'_{\alpha\beta} \cap D'_{\delta\varepsilon}, \ D'_{\alpha\beta} \cap D'_{\gamma\varepsilon}$$

are mapped, respectively, to $0, 1, \infty$.

For a fixed permutation $(\alpha, \beta, \gamma, \delta, \varepsilon)$ of $(0, 1, 2, 3, 4)$ we define

$$\pi'_\alpha : X' \to \mathbb{P}_1$$

by

$$\pi'_\alpha(z_0, z_1, z_2, z_3, z_4) = V(z_\beta, z_\gamma, z_\delta, z_\varepsilon) = \frac{(z_\beta - z_\varepsilon)(z_\gamma - z_\delta)}{(z_\beta - z_\delta)(z_\gamma - z_\varepsilon)} \quad (7.3)$$

and

$$\pi'_\beta : X' \to \mathbb{P}_1$$

by

$$\pi'_\beta(z_0, z_1, z_2, z_3, z_4) = V(z_\alpha, z_\gamma, z_\delta, z_\varepsilon) = \frac{(z_\alpha - z_\varepsilon)(z_\gamma - z_\delta)}{(z_\alpha - z_\delta)(z_\gamma - z_\varepsilon)}. \quad (7.4)$$

The images of

$$D'_{\gamma\delta} \cup D'_{\beta\varepsilon}, \ D'_{\delta\varepsilon} \cup D'_{\beta\gamma}, \ D'_{\gamma\varepsilon} \cup D'_{\beta\delta}$$

under π'_α are, respectively, 0, 1, ∞; and the images of

$$D'_{\gamma\delta} \cup D'_{\alpha\varepsilon}, \ D'_{\delta\varepsilon} \cup D'_{\alpha\gamma}, \ D'_{\gamma\varepsilon} \cup D'_{\alpha\delta}$$

under π'_β are also 0, 1, ∞. For any $t \in \mathbb{P}_1 \setminus \{0, 1, \infty\}$, the inverse image of t under π'_α or π'_β is a smooth projective line in X'. Hence, as with the $D_{\alpha\beta}$, the map

$$\pi' = \pi'_\alpha \times \pi'_\beta : X' \to \mathbb{P}_1 \times \mathbb{P}_1$$

is a blow-down of X' to $\mathbb{P}_1 \times \mathbb{P}_1$, collapsing $D'_{\gamma\delta}$, $D'_{\delta\varepsilon}$, $D'_{\gamma\varepsilon}$ to $(0, 0)$, $(1, 1)$, (∞, ∞), where $D'_{\alpha\varepsilon}$, $D'_{\alpha\gamma}$, $D'_{\alpha\delta}$ are mapped to $\mathbb{P}_1 \times \{0\}$, $\mathbb{P}_1 \times \{1\}$, $\mathbb{P}_1 \times \{\infty\}$ and $D'_{\beta\varepsilon}$, $D'_{\beta\gamma}$, $D'_{\beta\delta}$ are mapped to $\{0\} \times \mathbb{P}_1$, $\{1\} \times \mathbb{P}_1$, $\{\infty\} \times \mathbb{P}_1$. The image of $D'_{\alpha\beta}$ under π' is the diagonal of $\mathbb{P}_1 \times \mathbb{P}_1$. As the three curves $D'_{\gamma\delta}$, $D'_{\delta\varepsilon}$, $D'_{\gamma\varepsilon}$ are blown down to points of $\mathbb{P}_1 \times \mathbb{P}_1$, it follows that X' is compact and that these curves are exceptional (see [65]). We see that both X and X' can be obtained, up to isomorphism, as the blow-up of $\mathbb{P}_1 \times \mathbb{P}_1$ at the three points $(0, 0)$, $(1, 1)$, (∞, ∞). Therefore, X and X' are isomorphic as algebraic surfaces. By symmetry under the S_5 action, all ten curves $D'_{\alpha\beta}$ are exceptional and thus each has self-intersection -1. This configuration of ten curves has exactly the same intersection properties as the configuration of 10 curves $D_{\alpha\beta}$ on X. For any choice of permutation $(\alpha, \beta, \gamma, \delta, \varepsilon)$ of $(0, 1, 2, 3, 4)$ we also have the hexagon picture for X'. We have projective planes A' and B' where the lines through the blown-down $D'_{\alpha\beta}$ are the fibers of π'_α and π'_β respectively. The regular fibers are lines because their self-intersection numbers are each 1. There is an identification of X and X' that sends the $D_{\alpha\beta}$ to the $D'_{\alpha\beta}$. Moreover, the π_α, π_β go over to π'_α, π'_β, for every permutation $(\alpha, \beta, \gamma, \delta, \varepsilon)$ of $(0,1,2,3,4)$.

Definition 7.2 *We define Q' to be the open set in X' given by*

$$Q' = \{(z_0, z_1, z_2, z_3, z_4) \in \mathbb{P}_1^5 : z_i \neq z_j,\ i \neq j\}/\mathrm{Aut}(\mathbb{P}_1).$$

This is the configuration space of five points on the complex projective line. The action of S_5 on X' restricts to a well-defined action on Q'.

Fix a permutation $(\alpha, \beta, \gamma, \delta, \varepsilon)$ of $(0, 1, 2, 3, 4)$. Using the diagonal action of $\mathrm{Aut}(\mathbb{P}_1)$, we can apply the transformation given by $V(z, z_\gamma, z_\delta, z_\varepsilon)$ to normalize $z_\gamma, z_\delta, z_\varepsilon$ to $1, \infty, 0$. Upon setting $x = V(z_\alpha)$ and $y = V(z_\beta)$, we obtain the isomorphic space Q as follows.

Definition 7.3 *Let*

$$Q = \{(x, y) \in \mathbb{P}_1^2 : x, y \neq 0, 1, \infty,\ x \neq y\}.$$

We call Q the space of regular points.

The isomorphism from Q to Q' depends on the permutation of $(0, 1, 2, 3, 4)$ chosen and on the choice of cross ratio.

In Chapter 3, §3.3, we gave examples of finite covers $\pi : Y \to X$ of degree N with ramification along the ten lines of the blown-up complete quadrilateral configuration. The ramification indices were given in terms of five parameters,

$$0 < \mu_0, \ldots, \mu_4 < 1, \qquad \sum_{\alpha=0}^{4} \mu_\alpha = 2.$$

Recall that we assigned the ramifications

$$1 \le (1 - \mu_0 - \mu_\alpha)^{-1}, \qquad \alpha = 1, 2, 3, 4$$

to the four blown-up points, or exceptional curves, E_α, and we assigned the ramifications

$$1 \le (1 - \mu_\gamma - \mu_\delta)^{-1}, \qquad \gamma \neq \delta, \quad \gamma, \delta = 1, 2, 3, 4$$

to the proper transforms of the lines of the complete quadrilateral. We supposed, moreover, that

$$(1 - \mu_\alpha - \mu_\beta)^{-1} \in \mathbb{Z}, \qquad \alpha \neq \beta, \qquad \text{INT}$$

and we verified that the above conditions imply

$$\mathrm{Prop}(Y) = (3c_2(Y) - c_1^2(Y))/N = 0.$$

Hence, by the results of Chapter 6, the surface Y is a complex ball quotient B_2/Γ, where

$$B_2 = \{(u_0 : u_1 : u_2) : |u_1|^2 + |u_2|^2 < |u_0|^2\}$$

and $\Gamma \subset PU(2, 1) = \mathrm{Aut}(B_2)$ acts properly and discontinuously without fixed points. Up to permutation, there are eight quintuples $\mu_0, \ldots, \mu_4$ satisfying the above conditions (see Chapter 3, §3.3). In Chapter 5, we also considered the cases where Prop vanishes for negative or infinite weights. This yields a total of twenty-seven cases. Recall that in order for Prop to vanish when we allow a negative weight, the corresponding exceptional curve must be blown down. In the remainder of this chapter, we use the description of X as X', in order to relate the groups with discontinous action on B_2 to the monodromy groups of certain Appell hypergeometric functions.

7.2 APPELL HYPERGEOMETRIC FUNCTIONS

In this section we take a function-theoretic viewpoint. The space Q of regular points of the Appell hypergeometric function of Definition 1.3 plays the role that the space $\mathbb{P}_1 \setminus \{0, 1, \infty\}$ did for the Gauss hypergeometric function. Like the Gauss hypergeometric function, the Appell function arises as a solution of differential equations. Namely, it satisfies a system of partial differential equations, regular on Q. For complex numbers a, b, b', c, with $c \neq 0, -1, -2$, these equations are:

$$x(1-x)\frac{\partial^2 v}{\partial x^2} + y(1-x)\frac{\partial^2 v}{\partial x \partial y} + (c - (a+b+1)x)\frac{\partial v}{\partial x} - by\frac{\partial v}{\partial y} - abv = 0$$

$$y(1-y)\frac{\partial^2 v}{\partial y^2} + x(1-y)\frac{\partial^2 v}{\partial x \partial y} + (c - (a+b'+1)y)\frac{\partial v}{\partial y} - b'x\frac{\partial v}{\partial x} - ab'v = 0.$$

$$(7.5)$$

From these two partial differential equations we can derive a third one, namely,

$$(x - y)\frac{\partial^2 v}{\partial x \partial y} - b'\frac{\partial v}{\partial x} + b\frac{\partial v}{\partial y} = 0,$$

so that the solution space has dimension 3, and not dimension 4 as we may have expected (see [145]). We call the above system of partial differential equations the Appell system. It has regular singularities along the lines excluded by the definition of Q, that is, along

$$x = y, \qquad x, y = 0, 1, \infty.$$

Recall that the regular singular points of the Gauss system presented in Chapter 2 are at 0, 1, ∞. For a definition of regular singularities sufficient for our purposes, see [38].

Let (x_0, y_0) be a fixed point in Q. For $j = 0, 1, \infty, x_0, y_0$, we choose a closed loop in Q by starting at (x_0, y_0), and then by letting either x or y travel in the region $\mathbb{P}_1 \setminus \{0, 1, \infty, x_0, y_0\}$ up to j, by going once around j in the positive direction, and then by traveling back to x_0 or y_0. Five such loops are sufficient to generate $\pi_1(Q, (x_0, y_0))$ (see [145], p. 147). Let C be a closed loop in Q beginning and ending at (x_0, y_0). Let y_1, y_2, y_3 be three linearly independent solutions of Appell's system of partial differential equations in a neighborhood of (x_0, y_0). If we analytically continue these solutions around C, they remain linearly independent. Since they span the solution space, there is a nonsingular 3×3 matrix $M(C)$ such that ${}^t(y_1, y_2, y_3)$ becomes $M(C)^t(y_1, y_2, y_3)$ upon analytic continuation around C. A different choice of base point gives rise to a matrix conjugate to $M(C)$ in $GL_3(\mathbb{C})$. If $C_1 \circ C_2$ denotes the composition of two closed loops with base point (x_0, y_0), we have $M(C_1 \circ C_2) = M(C_1)M(C_2)$. If C_1 can be continuously deformed in Q to C_2 then $M(C_1) = M(C_2)$. From these remarks, we see that there is a homomorphism

$$M : \pi_1(Q, (x_0, y_0)) \to GL_3(\mathbb{C})$$

called the *monodromy representation* of the Appell system. The *monodromy group* is the image of the monodromy representation, and the *projective monodromy group* is defined as the image of the monodromy group under the natural map from $GL_3(\mathbb{C})$ to $PGL_3(\mathbb{C})$. If we change the base point (x_0, y_0) or the choice of basis for the solution space to Appell's system, then we conjugate M by an element of $GL_3(\mathbb{C})$. The conjugacy class of the monodromy group and of the projective monodromy group are uniquely determined by the differential equations.

In the region $|x| < 1, |y| < 1$, the following series is a solution to the Appell system:

$$F = F(a, b, b', c; x, y) = \sum_{m \geq 0, n \geq 0} \frac{(a, m+n)(b, m)(b', n)}{(c, m+n)} \frac{x^m y^n}{m! \, n!}, \quad (7.6)$$

where, for any complex number w and positive integer k, we have

$$(w, k) = w(w+1)\dots(w+k-1),$$

and we set $(w, 0) = 1$. This can be verified directly by setting

$$A_{m,n} = \frac{(a, m+n)(b, m)(b', n)}{(c, m+n)m!n!}$$

and observing that

$$f(m, n) =: \frac{A_{m+1,n}}{A_{m,n}} = \frac{(a + m + n)(b + m)}{(c + m + n)(1 + m)},$$

$$g(m, n) =: \frac{A_{m,n+1}}{A_{m,n}} = \frac{(a + m + n)(b' + n)}{(c + m + n)(1 + n)}. \tag{7.7}$$

The equations (7.5) defining the Appell system can be written in terms of the operators $D = x\frac{\partial}{\partial x}$ and $D' = y\frac{\partial}{\partial y}$ as follows:

$$\{(a + D + D')(b + D) - (c + D + D')(1 + D)x^{-1}\}v = 0,$$

$$\{(a + D + D')(b' + D') - (c + D + D')(1 + D')y^{-1}\}v = 0. \tag{7.8}$$

The verification is now easily completed by substituting

$$F = \sum_{m \geq 0, n \geq 0} A_{m,n} x^m y^n$$

in (7.8), using the expressions for $f(m, n)$ and $g(m, n)$ in (7.7), and noting that $Dx^m = mx^m$ and $D'y^n = ny^n$. Recall that

$$(1 - tw)^{-\beta} = \sum_{k \geq 0} \frac{(\beta, k)}{(1, k)} t^k w^k, \quad |w| < 1$$

and

$$\int_0^1 t^{\gamma-1}(1 - t)^{\delta-1} dt = \frac{\Gamma(\gamma)\Gamma(\delta)}{\Gamma(\gamma + \delta)},$$

for $\text{Re}(\gamma)$, $\text{Re}(\delta) > 0$. Therefore, for $|x|, |y| < 1$ and $\text{Re}(c) > \text{Re}(a) > 0$, we have

$$F = \frac{\Gamma(c)}{\Gamma(a)\Gamma(c - a)} \int_0^1 t^{a-1}(1 - t)^{c-a-1}(1 - tx)^{-b}(1 - ty)^{-b'} dt. \tag{7.9}$$

Setting $y = 0$ in (7.6) recovers the Gauss hypergeometric function $F(a, b, c; x)$ as a series. Setting $y = 0$ in (7.9) recovers the integral expression for that function.

The parameters a, b, b', c can be replaced by the $\mu_\alpha, \alpha = 0, \ldots, 4$, given by

$$\mu_0 = c - b - b', \quad \mu_1 = 1 - c + a, \quad \mu_2 = b, \quad \mu_3 = b', \quad \mu_4 = 1 - a \quad \text{(Par)}.$$

Note that $\sum_{\alpha=0}^{4} \mu_\alpha = 2$. Changing variables in the integrand by setting $t = \frac{1}{u}$, we obtain

$$F = \frac{\Gamma(2 - \mu_1 - \mu_4)}{\Gamma(1 - \mu_1)\Gamma(1 - \mu_4)} \int_1^\infty u^{-\mu_0}(u - 1)^{-\mu_1}(u - x)^{-\mu_2}(u - y)^{-\mu_3} du.$$

Suppose that $0 < \mu_\alpha < 1$. The analogous condition in the one-dimensional case implies, in particular, that the triangle given by the image of the ratio of two independent solutions of the Gauss hypergeometric equation has angle sum less than 1, and so is hyperbolic (see Chapter 2). Other solutions of the Appell system are given by

$$\int_g^h \omega(x, y),$$

for $g, h \in \{0, 1, \infty, x, y\}$, where

$$\omega(x, y) = u^{-\mu_0}(u - 1)^{-\mu_1}(u - x)^{-\mu_2}(u - y)^{-\mu_3} du.$$

A detailed proof of these facts can be found in [33]. The Euler integrals above determine a multivalued map from Q to a domain in the complex 2-ball (see Theorem 7.4 below).

Consider the action of S_5 on X' introduced above. In terms of coordinates $(z_0, \ldots, z_4) \in \mathbb{P}_1^5$, the hypergeometric integrals have the more symmetric form

$$\int_{z_\alpha}^{z_\beta} (u - z_0)^{-\mu_0}(u - z_1)^{-\mu_1}(u - z_2)^{-\mu_2}(u - z_3)^{-\mu_3}(u - z_4)^{-\mu_4} du.$$

We see that the integral expression for F is associated to a choice modulo the S_5–symmetry and modulo the normalization by $\mathrm{Aut}(\mathbb{P}_1)$. The other integral solutions correspond to other possible choices. The Appell system has rank 3. With the above assumptions, around each point $(x, y) \in Q$ we have the three linearly independent solutions

$$\int_1^\infty \omega(x, y), \quad \int_0^x \omega(x, y), \quad \int_0^y \omega(x, y).$$

Making the change of variable $u = xv$, we see that

$$\int_0^x \omega(x, y) = x^{1 - \mu_0 - \mu_2} \int_0^1 v^{-\mu_0}(xv - 1)^{-\mu_1}(v - 1)^{-\mu_2}(xv - y)^{-\mu_3} dv.$$

This integral therefore branches as $x^{1-\mu_0-\mu_2}$ along $x = 0$, $y \neq 0$. At the point $x = y = 0$, we see the branching exponent $1 - \mu_0 - \mu_2 - \mu_3 = \mu_1 + \mu_4 - 1$. Transferring this to the space X' we see that, along the open subset of $D'_{\alpha\beta}$ that does not intersect another $D'_{\gamma\delta}$, the hypergeometric integral

$$\int_{z_\alpha}^{z_\beta} (u - z_0)^{-\mu_0}(u - z_1)^{-\mu_1}(u - z_2)^{-\mu_2}(u - z_3)^{-\mu_3}(u - z_4)^{-\mu_4} du$$

will branch as $(z_\alpha - z_\beta)^{1 - \mu_\alpha - \mu_\beta}$. This is the origin of the integrality condition INT of §7.1. If $\mu_\delta + \mu_\varepsilon > 1$ and $D'_{\delta\varepsilon}$ is blown down, it becomes a triple intersection point with associated branching exponent $1 - \mu_\alpha - \mu_\beta - \mu_\gamma > 0$. If $(1 - \mu_\delta - \mu_\varepsilon)^{-1}$ is a negative integer, then $(1 - \mu_\alpha - \mu_\beta - \mu_\gamma)^{-1}$ is the integer given by its absolute value.

The above remarks give some intuition behind Theorem 7.4 below. It was first stated, without a rigorous proof, by Picard [116]. It was proved in full by Deligne and Mostow [33] and also studied, using other techniques, by Terada [136].

We define a ball quintuple to be a quintuple of real numbers $\mu = (\mu_\alpha)_{\alpha=0}^4$ with

$$0 < \mu_\alpha < 1, \quad \alpha = 0, \ldots, 4, \quad \sum_{\alpha=0}^4 \mu_\alpha = 2.$$

(Such μ are called disk quintuples in [109]). A point $z = (z_\alpha)_{\alpha=0}^4 \in \mathbb{P}_1^5$ is called μ-stable if, for all $x \in \mathbb{P}_1$, we have $\sum_{z_\alpha = x} \mu_\alpha < 1$. It is called μ-semistable if, for all $x \in \mathbb{P}_1$, we have $\sum_{z_\alpha = x} \mu_\alpha \leq 1$. Let M_{st} be the set of all μ-stable points, let M_{sst} be the set of all μ-semistable points, and let $M_{cusp} = M_{sst} \setminus M_{st}$. Let $\{I, J\}$ be a partition of $\{0, 1, 2, 3, 4\}$, that is, $I \cup J = \{0, 1, 2, 3, 4\}$ and $I \cap J = \phi$. If $\sum_{i \in I} \mu_i = \sum_{j \in J} \mu_j = 1$, then all points $z \in \mathbb{P}_1^5$ with components z_i equal for $i \in I$, or z_j equal for $j \in J$, but with $z_i \neq z_j$, for $i \in I$, $j \in J$, are in M_{cusp}. Conversely, all points in M_{cusp} are obtained in this way, each from a unique partition. As in [33], §4.1, we can define an equivalence relation $\mathcal{R}$ on M_{sst} as follows: we have $z \equiv z' \mod \mathcal{R}$ either when $z, z' \in M_{st}$ and there is a $\gamma \in \mathrm{Aut}(\mathbb{P}_1)$ acting diagonally such that $z' = \gamma(z)$, or when $z, z' \in M_{cusp}$ and the partitions of $\{0, 1, 2, 3, 4\}$ corresponding to z and z' coincide. Consider the quotient spaces

$$Q_{sst} = M_{sst}/\mathcal{R}, \quad Q_{st} = M_{st}/\mathcal{R}, \quad Q_{cusp} = M_{cusp}/\mathcal{R},$$

each with its quotient topology. The elements of Q_{cusp} are uniquely determined by their partitions. Therefore, Q_{cusp} is a finite set. The space Q_{sst} is Hausdorff, is compact, and can be given the structure of an algebraic variety (see [33],§4). The space Q can be realized as a subspace of Q_{sst} using the diagonal action of $\mathrm{Aut}(\mathbb{P}_1)$ on $\mathbb{P}_1^5$, and Q_{sst} is, in fact, a compactification of Q.

Given a quintuple $\mu = \{\mu_\alpha\}_{\alpha=0}^4$ of real numbers with $\sum_{\alpha=0}^4 \mu_\alpha = 2$, let Γ_μ be the monodromy group of Appell's system (7.5) with parameters related to

μ as in formula (Par) above. We now state the theorem of Picard, Terada, and Deligne–Mostow (see [33], Theorems 10.19 and 11.4).

Theorem 7.4 *Let $\mu = \{\mu_\alpha\}_{\alpha=0}^4$ be a ball quintuple. If, for all $\alpha \neq \beta$, we have*

$$(1 - \mu_\alpha - \mu_\beta)^{-1} \in \mathbb{Z} \cup \{\infty\} \qquad \text{INT},$$

then the monodromy group Γ_μ of Appell's system acts discontinuously on the complex 2-ball B_2 with finite covolume. The quotient space can be compactified by adding the finitely many points given by the orbits of the cusps of Γ_μ. The resulting projective variety is isomorphic to Q_{sst}.

The group Γ_μ of the theorem is a subgroup of $\text{Aut}(B_2) = \text{PU}(2, 1)$ acting by fractional linear transformations. We say that Γ_μ has finite covolume if the quotient space $\Gamma_\mu \backslash B_2$ has finite volume, and that it is cocompact if the quotient space is compact. When Γ_μ is cocompact, Q_{st} and Q_{sst} coincide and we have isomorphisms

$$Q_{st} \simeq \Gamma_\mu \backslash B_2 \simeq X_{st},$$

where X_{st} is X' with the exceptional curves $D'_{\alpha\beta}$ with $\mu_\alpha + \mu_\beta > 1$ blown down.

In their papers, Deligne and Mostow refer to lattices in $\text{PU}(1, 2)$. A lattice Γ in $\text{PU}(1, 2)$ is a discrete subgroup of $\text{PU}(1, 2)$ such that the quotient group $\Gamma\backslash\text{PU}(1, 2)$ has finite Haar measure. A subgroup of $\text{PU}(1, 2)$ that acts discontinuously on B_2 is a discrete subgroup of $\text{PU}(1, 2)$. Conversely, as $\text{PU}(1, 2)$ acts transitively on B_2 with compact isotropy group, a discrete subgroup of $\text{PU}(1, 2)$ acts discontinuously on B_2. By [109], Proposition 5.3, if the monodromy group Γ_μ associated to a ball quintuple μ is a discrete subgroup of $\text{PU}(1, 2)$, then it is a lattice in $\text{PU}(1, 2)$. Therefore the Γ_μ acting discontinuously on B_2 are precisely the lattices in $\text{PU}(1, 2)$, and they will have finite covolume.

The list of the twenty-seven quintuples μ, up to permutation of the μ_α, satisfying INT is given in [33], p. 86. The INT condition as given in [33] appears at first to be weaker than that given in Theorem 7.4, but it is in fact equivalent to it. Indeed, in [33] the INT condition for a ball quintuple is that for all $\alpha \neq \beta$ with $\mu_\alpha + \mu_\beta < 1$, we have $(1 - \mu_\alpha - \mu_\beta)^{-1} \in \mathbb{Z}$. One can check directly from the list of μ satisfying the INT condition of [33] that, for all of them, $(1 - \mu_\alpha - \mu_\beta)^{-1} \in \mathbb{Z}$ also when $\mu_\alpha + \mu_\beta > 1$. This list in [33], p. 86 corresponds to the first twenty-seven entries in Table 5.13, Chapter 5 (Le Vavasseur's list) with $\mu_i = \alpha_i/d$, $i = 0, \ldots, 4$, and the column DM of that table gives the corresponding entry in the Deligne–Mostow list. This list also coincides with the first twenty-seven entries of Table 5.12, Chapter 5. There,

$n_{\alpha\beta} = (1 - \mu_\alpha - \mu_\beta)^{-1}$, $\alpha \neq \beta$ (number 31 does not correspond to a ball quintuple).

Mostow [108] showed that Γ_μ acts discontinuously on B_2 with finite covolume also when the ball quintuple μ satisfies a weaker condition ΣINT. One can check easily that the ΣINT condition of [108] is equivalent to the following: for all $\alpha \neq \beta$ with $\mu_\alpha + \mu_\beta < 1$, we have

$$(1 - \mu_\alpha - \mu_\beta)^{-1} \in \begin{cases} \mathbb{Z} \cup \infty, & \mu_\alpha \neq \mu_\beta, \\ \frac{1}{2}\mathbb{Z} \cup \infty, & \mu_\alpha = \mu_\beta. \end{cases}$$

It is possible to work with the ΣINT condition because the permutation group S_5 acts biholomorphically on the configuration space X' defined at the beginning of the present chapter. This allows us to form the quotient of the space Q_{st} by the subgroup of S_5 leaving the ordered set of μ_αs invariant. This space is then a quotient of the ball by Γ_μ. For details, see [34] and [29], [30].

In [109], pp. 584–586, Mostow gives the list (calculated by computer by Thurston) of all ball quintuples satisfying ΣINT. It turns out there are fifty-four ball quintuples (up to permutation) satisfying ΣINT.

In [109], Mostow showed that there are nine ball quintuples μ that do *not* satisfy ΣINT but for which Γ_μ acts discontinuously on B_2 with finite covolume. In [127], Theorems 3.1 and 4.1, Sauter proved a conjecture of Mostow to the effect that these nine exceptional Γ_μ are commensurable to a Γ_ν where ν is a ball quintuple satisfying ΣINT. Recall that two groups Γ and Γ' are commensurable if their intersection $\Gamma \cap \Gamma'$ is of finite index in both Γ and Γ'.

In Chapter 2, §2.6, we introduced the notion of a ball quadruple. This is a quadruple $\mu = \{\mu_\alpha\}_{\alpha=0}^3$ of real numbers satisfying

$$0 < \mu_\alpha < 1, \quad \alpha = 0, \ldots, 3, \quad \sum_{\alpha=0}^3 \mu_\alpha = 2.$$

Let Γ_μ be the monodromy group of the Gauss hypergeometric differential equation of Chapter 2, (1.1), where $\mu_0 = c - b$, $\mu_1 = 1 + a - c$, $\mu_2 = b$, $\mu_3 = 1 - a$. Then, if for all $\alpha \neq \beta$, $(1 - \mu_\alpha - \mu_\beta)^{-1} \in \mathbb{Z} \cup \{\infty\}$, then Γ_μ is a hyperbolic triangle group acting discontinuously on the unit disk with finite covolume (see Chapter 2, §2.7). There are infinitely many such μ.

Deligne and Mostow [33], [108], [109] develop analogous criteria for the monodromy groups of the n-variable Lauricella hypergeometric functions to act discontinuously with finite covolume on the complex n-ball, $n \geq 1$.

We end this section with some remarks about the relation between some of the quintuples satisfying INT and the Ceva(r) arrangement; see also

Chapter 5, §5.5. As we saw in that same section, the diophantine condition

$$\frac{r}{n} + \frac{2}{m} = r - 2, \quad n \geq 2, \, r \geq 3$$

has only finitely many solutions, given in Table 5.1, Chapter 5. This condition refers to r lines passing through a point, with ramification n along each line and m at the intersection point. There are two such cases with $m = n$ and $n \geq 3$, namely $r = 4, n = m = 3$ and $r = 3, n = m = 5$. The Ceva(r) arrangement, $r \geq 2$, has r-fold points and triple points only. It can be obtained from the "inside" lines

$$(x_0 - x_1)(x_0 - x_2)(x_1 - x_2) = 0$$

of the $\overline{\text{Ceva}}(1)$ arrangement

$$x_0 x_1 x_2 (x_0 - x_1)(x_0 - x_2)(x_1 - x_2) = 0$$

by taking the r^2-fold covering of $\mathbf{P}_2$ defined by

$$\sqrt[r]{x_1/x_0}, \quad \sqrt[r]{x_2/x_0},$$

using the "outside" lines

$$x_0 x_1 x_2 = 0.$$

The three "inside" lines lift to the $3r$ lines of Ceva(r). Consider the following ramification data for Ceva(r): n along each of the $3r$ lines, m_3 over the threefold points and m_r over the r-fold points. Then, we obtain the following conditions for a corresponding covering of Ceva(r) to have vanishing proportionality:

$$\frac{3}{n} + \frac{2}{m_3} = 1, \quad \frac{r}{n} + \frac{2}{m_r} = r - 2.$$

(For Ceva(2) we have only the first equation.) Viewing this as a covering of $\overline{\text{Ceva}}(1)$, this corresponds to taking ramification r along the "outside" lines and n along the "inside" lines of $\overline{\text{Ceva}}(1)$. This gives the following values for the μs:

$$(\mu_0, \mu_1, \mu_2, \mu_3, \mu_4) = \left(\frac{1}{2} - \alpha, \frac{1}{2} - \alpha, \frac{1}{2} - \alpha, \frac{1}{2} + \alpha - \beta, \beta + 2\alpha \right),$$

with

$$2\alpha = \frac{1}{n}, \quad \beta = \frac{1}{r},$$

which is consistent with the INT condition. For $\beta = \frac{1}{2}$, we have $r = 2$ and $2\alpha = \frac{1}{n}$, with

$$(\mu_0, \mu_1, \mu_2, \mu_3, \mu_4) = \left(\frac{1}{2} - \alpha, \frac{1}{2} - \alpha, \frac{1}{2} - \alpha, \alpha, \frac{1}{2} + 2\alpha \right).$$

We obtain Ceva(2) from the inside lines of $\overline{\text{Ceva}}(1)$ via the fourfold cover of $\mathbb{P}_2$ defined above, with $r = 2$. The above values of the μs correspond to assigning the ramification n to each of the six lines of Ceva(2) and the value $m_3 = 2(1 - \frac{3}{n})^{-1}$ over the threefold points. The arrangement Ceva(2) is isomorphic to $\overline{\text{Ceva}}(1)$ and this last ramification data corresponds to the quintuple

$$\mu' = \left(\frac{1}{2} - \alpha, \frac{1}{2} - \alpha, \frac{1}{2} - \alpha, \frac{1}{2} - \alpha, 4\alpha \right),$$

which satisfies the INT condition for $n = 2, 3, 4, 5, 6, 9$. Therefore, the covering corresponding to μ is a fourfold cover of the one corresponding to μ'. Of the twenty-seven ball 5-tuples $\{\mu_i\}$ satisfying INT, there are twenty that are related to Ceva(r) for $r = 2, 3, 4, 5, 6, 7, 8$. There is also a case corresponding to $n = \infty$, with $\alpha = 0$, $\beta = \frac{1}{3}, \frac{1}{4}$, giving rise to a total of twenty-two cases with three or four of the μs equal.

7.3 ARITHMETIC MONODROMY GROUPS

Some of the monodromy groups of the Gauss and Appell systems provide examples of non-arithmetic groups acting on the complex ball. Traditionally, arithmetic groups have played the major role in number theory, but the non-arithmetic groups are also of interest. The quest to find new examples of non-arithmetic groups in higher dimensions has given birth to a great deal of interesting mathematics, including some of the work of Deligne and Mostow [33], [108] about monodromy groups of hypergeometric functions.

We recall some basic definitions ([18], [101], [106], §2, [126] and, especially, [33], §12). Let L be an algebraically closed field of characteristic 0 and let $\text{GL}_n(L)$ be the group of all invertible elements of $M_n(L)$, the set of $n \times n$ matrices with entries in L. Denote a matrix $g \in M_n(L)$ by $g = (g_{ij})$, where the $g_{ij} \in L$ are the matrix entries. The following embedding endows $\text{GL}_n(L)$ with the structure of an affine variety,

$$\text{GL}_n(L) \hookrightarrow L^{n^2+1},$$

$$g \mapsto (g_{ij}, (\det(g_{ij}))^{-1}),$$

having the ring of regular functions $R_L = L[X_{ij}, (\det(X_{ij}))^{-1}]$. An algebraic group of matrices is a subgroup G of $\mathrm{GL}_n(L)$ that is the zero set of an ideal of polynomials in R_L. We say G is defined over a subfield k of L if the ideal of all polynomials vanishing on G has generators in the ring $R_k = k[X_{ij}, (\det(X_{ij}))^{-1}]$.

Definition 7.5 *A linear algebraic group G is a subgroup of $\mathrm{GL}(V)$, for some finite dimensional vector space V over L, that can be identified with an algebraic group of matrices by choosing a basis for V over L. Suppose, in addition, that there is a vector space W, over a subfield k of L, such that $V = W \otimes_k L$. Then, G is said to be defined over k if, by choosing a basis for W over k, it can be identified with an algebraic group of matrices defined over k.*

A *morphism of algebraic groups* is a group morphism that induces a ring comorphism on the corresponding regular functions. A morphism of linear algebraic groups defined over k is itself defined over k if the induced comorphism on regular functions is defined over k. A linear algebraic group G defined over k acts on itself by inner automorphisms. For $g \in G$, the differential of

$$\mathrm{Int}(g) : x \to g \cdot x \cdot g^{-1}, \qquad x \in G,$$

at the identity $e \in G$ is denoted by $\mathrm{Ad}(g)$. Let $\mathcal{G}$ denote $\mathrm{Lie}(G)$, the Lie algebra of G. The map

$$g \mapsto \mathrm{Ad}(g)$$

is a k-morphism of G into $\mathrm{GL}(\mathcal{G})$, called the *adjoint representation of G*.

An algebraic set is *reducible* if it is the union of two proper closed subsets. It is *connected* if it is not the union of two proper disjoint closed subsets. An algebraic group is *irreducible* if and only if it is connected. The Zariski connected component of the identity will be denoted by G^z. The index of G^z in G is finite. If $L = \mathbb{C}$, every affine algebraic group G has the structure of a complex Lie group. Moreover, G is connected as an algebraic group if and only if it is connected as a Lie group. When G is defined over $\mathbb{R}$, its real points, $G(\mathbb{R})$, form a closed subgroup of $\mathrm{GL}_n(\mathbb{R})$ and, hence, a real Lie group. But, viewed as a Lie group, $G(\mathbb{R})$ need not be connected when G is connected as an algebraic $\mathbb{R}$-group, even though the Lie group has only finitely many connected components. The connected component of the identity with the usual topology on the Lie group is denoted by $G(\mathbb{R})^o$.

In characteristic 0, a semisimple algebraic group is called an *adjoint connected* group when it is connected and has trivial center [101], p. 43. Let G be a connected real semisimple Lie group with trivial center. Then, G

is the topological connected component $\mathbf{G}(\mathbb{R})^o$ of $\mathbf{G}(\mathbb{R})$, for some adjoint connected semisimple algebraic group $\mathbf{G}$ defined over $\mathbb{R}$ (see, for example, Proposition 3.6, p. 12, [126]).

Definition 7.6 *A subgroup Γ of G is arithmetic in G if and only if there exists an algebraic group A defined over $\mathbb{Q}$, a compact group K, and an analytic isomorphism ι of $A(\mathbb{R})^o$ onto $G \times K$ such that the projection of $\iota(A(\mathbb{Z}) \cap A(\mathbb{R})^o)$ onto G is commensurable with Γ.*

Suppose that G is a simple non-compact Lie group. Then, arguing as in [33], §12.2.4, we may suppose in the above definition that A is $\mathbb{Q}$-simple, that is, has only trivial Zariski closed normal subgroups over $\mathbb{Q}$. Therefore, A is of the form

$$A = \mathrm{Restr}_{F/\mathbb{Q}} B,$$

where B is an absolutely simple (i.e., $\overline{\mathbb{Q}}$-simple) algebraic group defined over a number field F. Here Restr denotes the restriction of scalars map (see [18], §7.16). We have

$$A(\mathbb{R}) = \prod_v B(F_v),$$

where the product is taken over the Archimedean places v of F, and F_v is the completion of F at v. The existence of ι as in Definition 3.2 means that for all the places v except one, say v_1, the group $B(F_v)$ is compact. Hence, in particular, v is a real place for $v \neq v_1$ (see [126], Prop. 3.5, p. 11). Moreover, $B(F_{v_1})^o$ is isomorphic to G, the isomorphism carrying $B(\mathcal{O})$ to a subgroup of G commensurable with Γ, where $\mathcal{O}$ is the ring of integers of F. If G is an absolutely simple Lie group, then v_1 is also real, so F is totally real. Identifying F as a subfield of $\mathbb{R}$ using v_1, the group B defines an F-structure on $\mathbf{G}$ if $G = \mathbf{G}(\mathbb{R})^o$.

The groups $\mathrm{PU}(1, n)$, for $n \geq 1$, are adjoint connected absolutely simple non-compact real Lie groups. We now impose this hypothesis on $G = \mathbf{G}(\mathbb{R})^o$. Suppose that $\mathbf{G}$ is defined over a number field F and that Γ is a subgroup of the F-rational points of G. Let k^{ad} be the trace field of Γ, that is, the field generated over $\mathbb{Q}$ by the elements of $\{\mathrm{Trace}(\mathrm{Ad}(\gamma)) \mid \gamma \in \Gamma\}$. Then, we have the following (see [33], Corollary 12.2.8):

Arithmeticity criterion: A lattice Γ in G is arithmetic in G if and only if, for each embedding σ of F in $\mathbb{R}$ not inducing the identity on k^{ad}, the group $^\sigma(\mathbf{G}_F)(\mathbb{R})$ is compact.

This arithmeticity criterion can be applied to the monodromy group of the Gauss and Appell hypergeometric differential equations, and reformulated in terms of the $\mu = \{\mu_\alpha\}$ (see [33], Proposition (12.7) and [109], and Proposition 5.4). For $n = 1, 2$, let $\mu_0, \ldots, \mu_{n+2}$ be rational numbers satisfying

$$0 < \mu_\alpha < 1, \quad \alpha = 0, \ldots, n + 1, \qquad \sum_\alpha \mu_\alpha = 2.$$

Let Γ_μ be the monodromy group of the corresponding hypergeometric differential equation(s). Let N be the least common denominator of the μ_α. For any $x \in \mathbb{R}$, let $\langle x \rangle$ denote the fractional part of x, that is, $0 \le \langle x \rangle < 1$ and $x - \langle x \rangle \in \mathbb{Z}$. Consider the set I of integers m coprime to N, with $1 < m < N - 1$, and let

$$s_m = \sum_\alpha \langle m\mu_\alpha \rangle.$$

Then Γ_μ is arithmetic in $PU(1, n)$ if and only if $s_m = 1$ or $n + 2$ for $m \in I$. When $n = 1$, the only possibilities for (s_m, s_{-m}) are $(0, 2)$, $(2, 0)$, $(1, 1)$. When $n = 2$, the only possibilities are $(0, 3)$, $(3, 0)$, $(1, 2)$, $(2, 1)$. The full list of arithmetic groups satisfying ΣINT when $n = 2$ is given in [109], pp. 584–586. There are thirty-nine such groups. Of the nine exceptional groups studied by Sauter, five are arithmetic. The property of arithmeticity is preserved under commensurability. Because the criterion in terms of the μ_α is easy to check, it is of interest to know which lattices in $PU(1, n)$ are commensurable with monodromy groups of hypergeometric functions. Such questions are studied in [34]. Similar arithmeticity criteria for the monodromy groups of the Lauricella hypergeometric functions of $n \ge 3$ complex variables are also developed in [33], [109].

We make some additional remarks about the case $n = 1$. A well-known example of an arithmetic group is given by the image of the modular group $SL(2, \mathbb{Z})$ in $PSL(2, \mathbb{R})$. Denote by $\Delta = \Delta(p, q, r)$ the conjugacy class in $PSL(2, \mathbb{R})$ of the Fuchsian triangle groups of signature (p, q, r) (for notation, see Chapter 2). The modular group is a representative of $\Delta(2, 3, \infty)$ and has generators

$$\begin{pmatrix} 0 & -1 \\ 1 & 0 \end{pmatrix}, \qquad \begin{pmatrix} 1 & 1 \\ 0 & 1 \end{pmatrix}.$$

However, the conjugacy class $\Delta = \Delta(2, 5, \infty)$ is non-arithmetic in $PSL(2)$ and has representative Hecke's triangle group G_5 generated by

$$\begin{pmatrix} 0 & -1 \\ 1 & 0 \end{pmatrix}, \qquad \begin{pmatrix} 1 & \frac{1}{2}(1 + \sqrt{5}) \\ 0 & 1 \end{pmatrix}.$$

The discussion in [33], §12, especially (12.2.6), gives a criterion, previously discovered by Takeuchi [134], for deciding whether or not a Fuchsian group of the first kind is arithmetic. A Fuchsian group of the first kind is a group of fractional linear transformations acting properly discontinuously on $\mathcal{H}$ with finite covolume.

Takeuchi's criterion: Let Γ be a Fuchsian group of the first kind, and let

$$T = \{\text{trace}(\gamma) : \gamma \in \Gamma\}.$$

Then Γ is arithmetic if and only if

  (i) $k = \mathbb{Q}(t)_{(t \in T)}$ is a number field,
  (ii) $T \subset \mathcal{O}_k$, where $\mathcal{O}_k$ is the ring of integers of k,
  (iii) if there is a Galois embedding $\sigma : T \hookrightarrow \mathbb{R}$ with $\sigma(t^2) \neq t^2$, for some $t \in T$, then $\sigma(T)$ is a bounded subset of $\mathbb{R}$.

In order to prove the criterion above, Takeuchi worked with the following definition of arithmeticity for Fuchsian groups of the first kind:

Definition 7.7 *A Fuchsian group of the first kind Γ is arithmetic in* PSL(2) *if and only if there is a quaternion algebra $\mathcal{B}$, defined over a totally real number field F, that is trivial at one infinite place v_1 of F and nontrivial at the other infinite places, and there is an order $\mathcal{O}$ in $\mathcal{B}$ with the following property. Viewing $\mathcal{B}$ as a subalgebra of $M_2(\mathbb{R})$ using a trivialization of $\mathcal{B}$ at the place v_1, the intersection of $\mathcal{O}$ with* SL(2, $\mathbb{R}$) *has image in* PSL(2, $\mathbb{R}$) *that is commensurable with Γ.*

It is well known, but not obvious, that for Fuchsian groups of the first kind Definitions 7.6 and 7.7 are equivalent. A full proof of this fact may be found in [106], §2.

Let Γ be a Fuchsian triangle group of signature (p, q, r). Takeuchi [134] showed that the field k occurring in (i), and generated over $\mathbb{Q}$ by the elements of T, is given by

$$k = \mathbb{Q}\left(\cos\left(\frac{\pi}{p}\right), \cos\left(\frac{\pi}{q}\right), \cos\left(\frac{\pi}{r}\right) \right).$$

For G_5, we have $(p, q, r) = (2, 5, \infty)$. Notice that

$$\begin{pmatrix} 1 & \frac{1}{2}(1 + \sqrt{5}) \\ 0 & 1 \end{pmatrix} \cdot \begin{pmatrix} 0 & -1 \\ 1 & 0 \end{pmatrix} = \begin{pmatrix} \frac{1}{2}(1 + \sqrt{5}) & -1 \\ 1 & 0 \end{pmatrix},$$

which has trace $t = \frac{1}{2}(1 + \sqrt{5})$. As $t^2 = \frac{1}{2}(3 + \sqrt{5})$, the Galois map of k sending $\sqrt{5}$ to $-\sqrt{5}$ does not leave t^2 fixed. Nonetheless, $\sigma(\mathrm{T})$ is unbounded. Therefore, this group is non-arithmetic.

In addition, Takeuchi [135] showed that there are, up to permutation, exactly eighty-five signatures (p, q, r) such that the triangle groups of signature (p, q, r) are arithmetic. Therefore, there are infinitely many signatures giving rise to non-arithmetic triangle groups.

7.4 SOME REMARKS ABOUT THE SIGNATURE

At this point, rather than consider the proportionality, it is interesting to study an invariant called the *signature*, which was introduced in Chapter 3,§3.4. Recall that, for a complex surface X, the signature, $\mathrm{sign}(X)$, of X is given by

$$3\mathrm{sign}(X) = (c_1^2 - 2c_2)(X). \tag{7.10}$$

Consider a good covering of degree N of a complex surface X by a complex surface Y (see Definition 1.1 of Chapter 3, §3.1). Then, with the notations of Chapter 3, and letting $x_i = 1 - \frac{1}{b_i}$, we have

$$\frac{3\mathrm{sign}(Y)}{N} = 3\mathrm{sign}(X) - \sum_i D_i \cdot D_i + \sum_i \frac{1}{b_i^2} D_i \cdot D_i$$

$$= 3\mathrm{sign}(X) - 2\sum_i x_i D_i \cdot D_i + \sum_i x_i^2 D_i \cdot D_i.$$

This formula can be derived using considerations similar to those employed in Chapter 3. In the example of the complete quadrilateral, we have ten divisors D_i on the blow-up $X = \widehat{\mathbb{P}}_2$ of $\mathbb{P}_2$. The indices i correspond to the ten unordered pairs $\{\alpha\beta\}$, where $0 \le \alpha \ne \beta \le 4$ as in §7.1. Each of these divisors has self-intersection -1. When all the $b_i > 0$, we therefore have

$$\frac{3\mathrm{sign}(Y)}{N} = 1 - \sum_{i=1}^{10} \frac{1}{b_i^2}. \tag{7.11}$$

We can extend this formula to the case where some of the b_i are negative or infinite, as in Chapter 5, §§ 5.4 and 5.6. When $b_i < 0$, we have N/b_i^2 exceptional curves on Y over D_i which we blow down. When $b_i = \infty$, we can take any ramification m_i along D_i. This gives rise to a union C_i of elliptic curves on Y with total self-intersection number $-N/m_i^2$. Let $C = \sum_i C_i$.

When we blow down an exceptional curve over a D_i with $b_i < 0$, we add 1 to c_1^2 and subtract 1 from c_2. Letting Y' be the surface obtained from Y by blowing down all such curves, we have, from (7.10) and (7.11),

$$\frac{3\text{sign}(Y')}{N} = \frac{3\text{sign}(Y)}{N} + \sum_{b_i<0} \frac{3}{b_i^2}$$

$$= 1 - \sum_{b_i=\infty} \frac{1}{m_i^2} - \sum_{b_i>0} \frac{1}{b_i^2} + \sum_{b_i<0} \frac{2}{b_i^2}$$

$$= 1 - \sum_{b_i>0} \frac{1}{b_i^2} + \sum_{b_i<0} \frac{2}{b_i^2} + \frac{C \cdot C}{N}. \tag{7.12}$$

Let Y'' be the surface obtained from Y' by removing the elliptic curves. If the set of elliptic curves is nonempty, then Y'' is not compact. In the notation of Chapter 5, §5.4, we define the "non-compact" signature $\overline{\text{sign}}(Y'')$ of Y'' by

$$3\overline{\text{sign}}(Y'') = c_1^2(Y', C) - 2c_2(Y', C).$$

Then, we have

$$\frac{3\overline{\text{sign}}(Y'')}{N} = \frac{3\text{sign}(Y'')}{N} - \frac{C \cdot C}{N}$$

$$= 1 - \sum_{i=1}^{10} \frac{a_i}{b_i^2}, \tag{7.13}$$

where $a_i = -2$ if $b_i < 0$ and $a_i = 1$ otherwise. Recall from §7.2 that there are twenty-seven quintuples $\mu = \{\mu_\alpha\}_{\alpha=0}^{4}$ with $0 < \mu_\alpha < 1$, $\sum_\alpha \mu_\alpha = 2$, that satisfy INT. They correspond to the first twenty-seven entries in Table 5.13 of Chapter 5. We have $b_{\alpha\beta} = (1 - \mu_\alpha - \mu_\beta)^{-1}, \alpha \neq \beta$. As explained in Chapter 5, §5.5, for these weights the proportionality $\text{Prop}(Y', C)$ vanishes. Therefore, denoting $c_2(Y', C)$ by $\overline{c}_2(Y'')$, we have

$$\frac{3\overline{\text{sign}}(Y'')}{N} = \frac{\overline{c}_2(Y'')}{N} = 1 - \sum_{\alpha,\beta} \frac{a_{\alpha\beta}}{b_{\alpha\beta}^2}. \tag{7.14}$$

As in Chapter 5, let X be the space obtained by blowing up the complete quadrilateral in the four triple points. Let X' be obtained from X by blowing down the divisors with negative weights, and let X'' be obtained from X' by contracting the divisors with infinite weights. Then X'' is a ball quotient $X'' = \Gamma_\mu \backslash B_2$, as shown in Chapter 6. We can define the volume of X'' to be

$$\text{vol}(X'') = \frac{\overline{c_2}(Y'')}{N}, \tag{7.15}$$

which we may view as a generalized "Gauss-Bonnet" formula. Of the twenty-seven quintuples satisfying **INT**, there are eighteen for which infinite ramification does not occur, and for all the quintuples negative ramification only occurs at four disjoint lines of the ten lines in $\widehat{\mathbb{P}}_2$. For example, the following 5-tuple has $\mu_0 + \mu_\alpha > 1, \alpha = 1, \dots 4$:

$$\left(\frac{10}{12}, \frac{5}{12}, \frac{3}{12}, \frac{3}{12}, \frac{3}{12} \right).$$

Consider once again the space X obtained by blowing up the four triple points of the complete quadrilateral. Assign the ramification 5 to the ten lines of the resulting configuration on X. The corresponding values of the μ_α are $\mu_\alpha = \frac{2}{5}$ for all $\alpha = 0, \dots, 4$. Then, from (4.6) and (4.7), with $X = \Gamma_\mu \backslash B_2$, we have

$$\text{vol}(X) = 1 - \frac{10}{25} = \frac{3}{5}$$

and

$$\text{Euler number}(Y) = (\text{degree of cover}) \times \frac{3}{5} = 3 \cdot 5^4.$$

Recall that $\mathbb{P}_1$ has many differing volumes, corresponding to differing triangle groups. Similarly, because X has many representations as a ball quotient, it has many different volumes. The ball quotients arising in this situation are, in general, of the form $X = \Gamma \backslash B_2$, where Γ does not operate freely. A "good" covering is obtained by taking a subgroup of finite index Γ' in Γ that does operate freely. According to results of Borel [18] and Selberg [129], such Γ' exist. This is a special case of the discussion of Chapter 6.

Appendix A

Torsion-Free Subgroups of Finite Index
by Hans-Christoph Im Hof

In Chapter 6, §6.2, we discussed the notion of b-space $(X; b)$ due to Kato. In [75], Kato considers conditions for (real) b-spaces $(X; b)$ to be (globally) uniformizable, that is, for there to be a transformation group (G, M) with $G \backslash M = (X; b)$, where the group G may be infinite. The topological space X is assumed to be a connected separable Hausdorff space, and M is assumed to be a connected paracompact manifold, while G is a group of topological transformations of X. Kato also assumes that the pair (G, M) is "proper and locally smooth", which places some natural conditions on the action of G and on its isotropy subgroups G_z, $z \in M$. If $\dim_{\mathbb{R}}(X) = n$, then $(X; b)$ is said to be a branchfold if $\dim_{\mathbb{R}}(\Sigma X) \leq (n - 2)$, where $\Sigma X = \Sigma(X; b)$ is the set of points of X where b is at least 2. Kato shows that there is a 1-1 correspondence between isomorphism classes of uniformizations of a branchfold $(X; b)$ and certain normal subgroups, called b-complete normal subgroups, of the fundamental group $\pi_1(X \setminus \Sigma X)$ of $X \setminus \Sigma X$. This answers a question of Fox [43], p. 252, for locally smooth regular branched coverings. If a uniformization (G, M) of $(X; b)$ exists, then the quotient map $\varphi : M \to X$ is a regular branched covering with deck transformation group G. That is, $\varphi_0 = \varphi \mid_{M_0} : M_0 \to X_0$ is a regular covering associated to a normal subgroup K of $\pi_1(X_0)$, where $X_0 = X \setminus \Sigma X$ such that $\pi_1(X_0)/K = G$, where $M_0 = M \setminus \varphi^{-1}(\Sigma X)$, so that (G, M) is associated to $K \subset \pi_1(X_0)$. Fox proved a converse whereby, for a normal subgroup K of $\pi_1(X_0)$, there is a unique regular branched covering $\varphi : M \to X$ with deck transformation group $G = \pi_1(X_0)/K$ that is a suitable completion of $\varphi_0 : M_0 \to X_0 \subset X$ associated to K.

In this book, we are interested in the complex b-spaces $(X_0''; b)$ associated to weighted line arrangements in the projective plane, where X_0'' is the space we derived from a weighted line arrangement in Chapter 6 and b reflects the positive weights on the blow-up of a line arrangement at its singular points. We found necessary and sufficient conditions on the weights for there to be a global cover of $B_2 \to \Gamma \backslash B_2 \simeq X_0''$ with ramification data matching that of b. What's more, we showed that the existence of such a cover of X_0'' was equivalent to equality in a suitable Miyaoka-Yau inequality that could be expressed in terms of the original weight data. In keeping with the spirit of Chapter 2, and in view of Kato's work referred to above, it is of interest to examine the work of Fox in real dimension 2 [42], which answered a conjecture of Fenchel.

When F. Hirzebruch gave the lectures at ETH Zürich that led to this book, Hans-Christoph Im Hof, of the University of Basel, Switzerland, was kind enough to agree to write an account of Fox's work. This account now follows.

We shall discuss a theorem asserting the existence of torsion-free subgroups of finite index in finitely generated Fuchsian groups. This assertion has been known as Fenchel's Conjecture since 1948 (Nielsen [115]; we have not found an earlier reference to the statement of the conjecture).

Below, we shall give the definition of the signature of a Fuchsian group (see, also, Chapter 2). It is not too difficult to prove the theorem for a Fuchsian group of signature $(g, n; p_1, \ldots, p_n)$, with $p_j = \infty$ for some j or with $g > 0$. Moreover, the remaining cases can be reduced to those of signature $(0, 3; p_1, p_2, p_3)$, p_1, p_2, p_3 finite, without too much difficulty. The difficult part is proving the theorem for cocompact triangle groups.

Apparently, the first proof for triangle groups was given by Fox [42] in 1952. Later, proofs by different methods were found by Macbeath [96], Feuer [41], Zieschang et al. [147], Mennicke [100], and Edmonds et al. [37]. However, in 1984 Brenner and Lyndon pointed out that the main step in Fox's proof was already carried out by G. A. Miller [102] in 1900.

Using more sophisticated methods, Selberg [129] and Borel [18] have derived properties of matrix groups which yield Fenchel's Conjecture as a corollary.

A.1 FUCHSIAN GROUPS

By definition, a Fuchsian group Γ is a discrete subgroup of $\mathrm{PSL}(2,\mathbb{R})$, the group of orientation preserving isometries of the hyperbolic plane. A finitely generated Fuchsian group can be presented by generators and relations in the following way:

$$\Gamma = < s_i, u_j, v_j; s_1^{p_1} = \ldots = s_m^{p_m} = s_1 \ldots s_n [u_1, v_1] \ldots [u_g, v_g] = 1 >,$$

with $2 \le p_1, \ldots, p_m < \infty$ and $p_{m+1} = \ldots = p_n = \infty$ for some $m \le n$. This presentation is generally not unique.

The set of numbers $(g, n; p_1, \ldots, p_n)$ is called the *signature* of the Fuchsian group. In most cases, the signature satisfies the hyperbolicity condition

$$\mu = 2(g - 1) + \sum_{i=1}^{n} \left(1 - \frac{1}{p_i}\right) > 0.$$

Conversely, an abstract group presented as above and satisfying the hyperbolicity condition can be realized as a discrete subgroup of $\mathrm{PSL}(2,\mathbb{R})$ with a fundamental domain of finite hyperbolic area $2\pi\mu$. Here, μ is the negative of the Euler volume. Therefore, these groups can be realized as Fuchsian groups with finite covolume.

If we retain the presentation above, but allow any value of μ, we obtain a slightly larger class of groups, including discrete groups of Euclidean isometries, when $\mu = 0$, or spherical isometries, when $\mu < 0$. The elementary Fuchsian groups are either cyclic or infinite dihedral, so they belong to the class of groups with $\mu \leq 0$.

Most of the proofs of Fenchel's Conjecture rely on the fact that, in a Fuchsian group, every element of finite order is conjugate to a power of one of its generators of finite order. This is also true in a discrete group of Euclidean or spherical isometries. This, in turn, follows from the fact that fixed points of elements of finite order lie on the boundary of a fundamental domain (see [98]).

A.2 FENCHEL'S CONJECTURE

Fenchel's Conjecture is sometimes stated as the existence of a torsion-free subgroup of finite index. Alternate formulations require that that subgroup also be normal. It is easy to show that the intersection of all the conjugates of a subgroup of finite index is a normal subgroup of finite index (see, for example, [24].) Therefore, as far as existence is concerned, the two versions of Fenchel's Conjecture are equivalent. However, for the question of minimizing the index of a torsion-free subgroup, the two versions differ. In Edmonds et al. [37], Fenchel's Conjecture is proved by giving necessary and sufficient conditions for a number to be the index of a torsion-free subgroup. The method of [37] does not apply to normal subgroups.

Here, we shall present a proof of Fenchel's Conjecture in which the desired subgroups appear as kernels of suitable homomorphisms. Therefore we shall be concerned mainly with normal subgroups.

In order to reformulate Fenchel's Conjecture, we introduce the notion of a conservative homomorphism.

Definition A.1 *A group homomorphism is called* conservative *if each element of finite order is mapped onto an element of the same order.*

Lemma A.2 *A group Γ has a torsion-free normal subgroup of finite index if and only if there is a conservative homomorphism of Γ onto a finite group.*

Proof.   (i) Let N be a torsion-free normal subgroup of finite index in Γ. We have to show that the quotient map $\varphi : \Gamma \to \Gamma/N$ is conservative. Let $g \in \Gamma$ be an element of finite order a. Clearly $\varphi(g)^a = \varphi(g^a) = 1$.

Assume $\varphi(g)^b = 1$ for a number b dividing a. Then $g^b \in N$ and $(g^b)^{a/b} = 1$. This is possible only for $b = a$.

(ii) Let $\varphi : \Gamma \to F$ be a conservative homomorphism of Γ onto a finite group. It is obvious that $\text{Ker}\varphi$ does not contain elements of finite order apart from the identity.

$\square$

A.3 REDUCTION TO TRIANGLE GROUPS

Let Γ be a group of signature $(g, n; p_1, \ldots, p_n)$, that is, a group with presentation

$$< s_i, u_j, v_j; s_1^{p_1} = \ldots = s_m^{p_m} = s_1 \ldots s_n[u_1, v_1] \ldots [u_g, v_g] = 1 >,$$

where $g \geq 0$, $n \geq 0$, $0 \leq m \leq n$, $2 \leq p_1, \ldots, p_m < \infty$, and $p_{m+1} = \ldots = p_n = \infty$.

First, we prove Fenchel's Conjecture for the cases $m < n$, which is the noncompact case, and $m = n$, $g > 0$, which is the compact case of positive genus. We essentially follow Edmonds et al. [37].

Let λ be the least common multiple of the orders $p_1, \ldots, p_m$, and hence of all orders of finite order elements of Γ (see §1). Let C_λ be the cyclic group $\mathbb{Z}/\lambda\mathbb{Z}$. Define $\varphi : \Gamma \to C_\lambda$ by setting

$$\varphi(s_i) = \lambda/p_i \qquad \text{for } i = 1, \ldots, m,$$

$$\varphi(s_{m+1}) = -\sum_{i=1}^{m} \lambda/p_i,$$

$$\varphi(s_j) = 0 \qquad \text{for } j = m+2, \ldots, n,$$

$$\varphi(u_k) = \varphi(v_k) = 0 \qquad \text{for } k = 1, \ldots, g.$$

This defines a conservative homomorphism.

Now, assume $m = n$ and $g > 0$. Let $D_{2\lambda}$ be the dihedral group of order 4λ with presentation

$$< x, y; x^{2\lambda} = y^2 = 1, \ yxy^{-1} = x^{-1} > .$$

Define $\psi : \Gamma \to D_{2\lambda}$ by setting

$$\psi(s_i) = x^{2\lambda/p_i},$$

$$\psi(u_1) = y,$$

$$\psi(v_1) = x^h, \qquad \text{where } h = \sum_{i=1}^{n} \lambda/p_i,$$

$$\psi(u_k) = \psi(v_k) = 1, \qquad \text{for } k = 2, \ldots, g.$$

Then, $\psi(s_1 \ldots s_n) = x^{2h}$ and $\psi([u_1, v_1]) = x^{-2h}$; thus, ψ is a homomorphism.

We claim that $N = \psi^{-1}(\{1, y\})$ is the desired subgroup. First, the group N has finite index 2λ in Γ. Next, it is torsion free, since an element of finite order in Γ is conjugate to a power of some s_i and is therefore mapped into $C_{2\lambda}$, the normal subgroup of $D_{2\lambda}$ generated by x.

We are left with groups of signature $(0, n; p_1, \ldots, p_n)$, such that $2 \leq p_i < \infty$ for all i. If $n \leq 2$, the corresponding groups are either trivial or cyclic of finite order, and Fenchel's Conjecture holds for trivial reasons. In the last reduction step, we may therefore restrict ourselves to the cases $n \geq 3$.

Assume the theorem is true for $n = 3$.

Consider Γ of signature $(0, n; p_1, \ldots, p_n)$, or $(p_1, \ldots, p_n)$ for short. Following Fox [42], we shall construct a conservative homomorphism of Γ into a finite group.

For $i = 2, \ldots, n - 1$, let Δ_i be the triangle group of signature (p_{i-1}, p_i, p_{i+1}), and again denote its generators by s_{i-1}, s_i, s_{i+1}. There are obvious homomorphisms from Γ to Δ_i that send $s_j \in \Gamma$ to $s_j \in \Delta_i$, for $j = i - 1, i, i + 1$, and to 1 otherwise.

By assumption, each Δ_i admits a conservative homomorphism $\varphi_i : \Delta_i \to F_i$ onto some finite group F_i. Let $\varphi_i(s_{i-1}) = a_{i-1}$, $\varphi_i(s_i) = b_i$, and $\varphi_i(s_{i+1}) = c_{i+1}$. Combine the homomorphisms $\varphi_2, \ldots, \varphi_{n-1}$ to obtain a map $\varphi : \Gamma \to F = F_2 \times \ldots \times F_{n-1}$ by setting

$$\varphi(s_1) = (a_1, 1, \ldots, 1),$$

$$\varphi(s_2) = (b_2, a_2, 1, \ldots, 1),$$

$$\varphi(s_i) = (1, \ldots, 1, c_i, b_i, a_i, 1, \ldots, 1),$$

$$\varphi(s_{n-1}) = (1, \ldots, 1, c_{n-1}, b_{n-1}),$$

$$\varphi(s_n) = (1, \ldots, 1, c_n).$$

It turns out that the order of $\varphi(s_i)$ is p_i. Any other element $g \in \Gamma$ of finite order is conjugate to some s_i^h; hence $\varphi(g)$ is conjugate to $\varphi(s_i^h)$. Since conjugation preserves the order of an element, $\varphi(g)$ has the same order as g. Therefore, φ is conservative.

A.4 TRIANGLE GROUPS

There are quite a number of proofs of Fenchel's Conjecture for triangle groups. We shall review below the different strategies of these proofs; one of them will be presented in more detail. The strategies (a) and (d) apply to all triangle groups. The other proofs work only in the hyperbolic case. However, for finite (i.e., spherical) triangle groups, Fenchel's Conjecture is obviously true, and it is a consequence of Bieberbach's Theorem for Euclidean triangle groups [8], [15].

Proof. (a) Proofs using permutation groups: Miller [102], Fox [42], and Brenner–Lyndon [22] prove the following lemma. □

Lemma A.3 *Let* $a, b, c \geq 2$ *be any three integers. Then, there are permutations* A *and* B *of a suitable finite set, such that* A *has order* a, B *has order* b, *and* $C = AB$ *has order* c.

The proofs of this lemma are elementary, but rather intricate. They involve many case-by-case arguments according to the respective sizes and parities of the numbers a, b, c.

Using Lemma A.3, a conservative homomorphism from the triangle group Δ of signature (a, b, c) into a suitable symmetric group is defined by mapping the generators of Δ onto the permutations A, B, C, respectively. □

Proof. (b) Proofs using Galois fields: Macbeath [96], Feuer [41], and Zieschang et al. [146] [147] prove Fenchel's Conjecture by mapping the triangle group into the group of special linear fractional transformations, PSL(2, q), with coefficients in a suitable Galois field $\mathbb{F}_q$. By choosing the prime power $q = p^n$ properly, this map can be turned into a conservative homomorphism. □

Proof. (c) Proofs Selberg [129] and Borel [18]: they prove the existence of a torsion-free normal subgroup of finite index in every finitely generated subgroup of GL(n, $\mathbb{C}$). This implies Fenchel's Conjecture for Fuchsian groups.

A proof given by Mennicke [100] is adapted to the particular case of a hyperbolic triangle group. Here, an embedding of the triangle group into the Lorentz group with coefficients in a certain number field is made conservative after projection modulo a properly chosen congruence subgroup.

More precisely, let Δ be a cocompact hyperbolic triangle group of signature (p_1, p_2, p_3). To the numbers p_i, we associate the numbers $c_i = \cos \frac{\pi}{p_i}$, $i = 1, 2, 3$. Let J denote the ring of integers of the algebraic number field $\mathbb{Q}(c_1, c_2, c_3)$. Clearly, $2c_1, 2c_2, 2c_3$ belong to J. We shall construct a faithful representation of Δ in the orthogonal group SO(2, 1; J). We work over the reals with a model of non-Euclidean geometry in $\mathbb{R}^{2,1}$, the metric having signature $+, +, -$. Passing to the projective space associated with $\mathbb{R}^{2,1}$, this gives a model of the hyperbolic plane as the interior of the unit circle with respect to this metric. In Chapter 1, we worked with the model of the hyperbolic plane as the interior of the unit disk.

Choose a basis $\{e_1, e_2, e_3\}$ of the Minkowski space $\mathbb{R}^{2,1}$ with corresponding Gram matrix

$$\begin{pmatrix} 1 & -c_3 & -c_2 \\ -c_3 & 1 & -c_1 \\ -c_2 & -c_1 & 1 \end{pmatrix}.$$

The vectors e_1, e_2, e_3 are the outer normals of a simplicial cone with dihedral angles $\frac{\pi}{p_1}, \frac{\pi}{p_2}, \frac{\pi}{p_3}$. In the projective space associated with $\mathbb{R}^{2,1}$, this cone gives rise to a hyperbolic triangle. The reflections at the facets of the cone, or, likewise, at the sides of the triangle, are described by

$$x \in \mathbb{R}^{2,1} \mapsto R_i = R_i(x) = x - 2(x, e_i)e_i, \qquad i = 1, 2, 3.$$

With respect to the basis $\{e_1, e_2, e_3\}$, we have the matrix representation

$$R_1 = \begin{pmatrix} -1 & 2c_3 & 2c_2 \\ 0 & 1 & 0 \\ 0 & 0 & 1 \end{pmatrix}, \quad R_2 = \begin{pmatrix} 1 & 0 & 0 \\ 2c_3 & -1 & 2c_1 \\ 0 & 0 & 1 \end{pmatrix},$$

$$R_3 = \begin{pmatrix} 1 & 0 & 0 \\ 0 & 1 & 0 \\ 2c_2 & 2c_1 & -1 \end{pmatrix},$$

which are elements of $O(2, 1; J)$.

The products $S_1 = R_2 R_3$, $S_2 = R_3 R_1$, and $S_3 = R_1 R_2$ are elements of the group $SO(2, 1; J)$. They describe rotations by the angles $\frac{2\pi}{p_1}, \frac{2\pi}{p_2}, \frac{2\pi}{p_3}$, and they satisfy the relations:

$$S_1^{p_1} = S_2^{p_2} = S_3^{p_3} = S_1 S_2 S_3 = \mathrm{Id}.$$

Mapping the generators s_1, s_2, s_3 of Δ onto the matrices S_1, S_2, S_3 defines a faithful representation $\rho : \Delta \to SO(2, 1; J)$.

Now, choose a prime ideal $\mathfrak{p} \subset J$. By composing ρ with the quotient map, we obtain $\varphi : \Delta \to SO(2, 1; J/\mathfrak{p})$. Observe that $J/\mathfrak{p}$, and hence $SO(2, 1; J/\mathfrak{p})$, are finite. To complete the proof, we have to show that $\mathfrak{p}$ may be chosen in such a way that

$$N = \mathrm{Ker}\, \varphi = \{g \in \Delta \mid \rho(g) \equiv \mathrm{Id}\ \mathrm{mod}\ \mathfrak{p}\}$$

does not contain elements of finite order.

Assume $g \in \Delta$ is a nontrivial element of finite order. Since g is conjugate to a power of a generator of Δ, it is a rotation by an angle of the form $\alpha = \frac{2\pi h}{p_i}$ for some $i = 1, 2, 3$ and $1 \leq h < p_i$. Therefore, $\mathrm{tr}(\rho(g)) = 1 + 2 \cos \alpha$. If $g \in N$, then $\rho(g) \equiv \mathrm{Id} \bmod \mathfrak{p}$; hence $\mathrm{tr}(\rho(g)) \equiv 3 \bmod \mathfrak{p}$. We therefore have

$$1 + 2 \cos \alpha \equiv 3 \bmod \mathfrak{p},$$

and $2(\cos \alpha - 1) \in \mathfrak{p}$.

It is always possible to choose $\mathfrak{p}$ in such a way that it avoids the finitely many numbers

$$2 \left(\cos \frac{2\pi h}{p_i} - 1 \right), \qquad i = 1, 2, 3 \text{ and } 1 \leq h < p_i.$$

□

Proof. (d) Proof using tessellations: In Edmonds et al. [37] it is shown that torsion-free subgroups of finite index in groups of signature $(p_1, \ldots, p_n)$ correspond to certain tessellations of closed orientable surfaces. In [36] and [37], explicit constructions of the desired tessellations are given. In addition, these methods determine which numbers occur as the indices of torsion-free subgroups of the given group that are not necessarily normal. □

Appendix B

Kummer Coverings

Consider an arrangement C of $k \geq 3$ lines $L_1, \ldots, L_k$ in $\mathbb{P}_2$ given by linear equations $l_1 = 0, \ldots, l_k = 0$. Suppose that the k lines do not all pass through the same point, that is, they do not form a pencil. Let X be the complex surface obtained by blowing up all the singular points of the arrangement. Let D_i be the proper transform of L_i, $i = 1, \ldots, k$. The number of blow-ups of $\mathbb{P}_2$, which are necessary to obtain X, is $s = \sum_{r \geq 3} t_r$ where, as before, we denote by t_r the number of points through which pass r lines of the arrangement. Along each D_i assign the ramification $n_i = n$, and along the exceptional divisors E_j, $j = 1, \ldots, s$, assign the ramification $m_j = n$. We shall now construct a good covering Y of X of degree $N = n^{k-1}$ (see Definition 5.1), with the ramifications $n_i = m_j = n$, $i \in I$, $j \in J$. The quotient l_i/l_j of two linear equations in homogeneous coordinates is a meromorphic function on X. We consider all the nth roots $\sqrt[n]{l_i/l_j}$, $i \neq j$. The covering Y is then defined by the property that these roots become univalued on Y. The surface Y is smooth and covers X with degree n^{k-1}. To see this, it is enough to consider the nth roots of l_i/l_k, $i = 1, \ldots, k-1$, for example. We have to check conditions (i) to (iii) of Definition 5.1. We only treat condition (ii), the others being easily verified. If we have a regular point at the intersection of two lines of the arrangement, say L_1 and L_2, and if L_3 is any other line, then $\sqrt[n]{l_1/l_3}$ and $\sqrt[n]{l_2/l_3}$ define a local covering of degree n^2. These roots can be used as local coordinates in Y. If we have a singular point of the arrangement of order r, through which $L_1, L_2, \ldots, L_r$ pass, and if L_{r+1} is any other line, then $\sqrt[n]{l_1/l_2}$ and $\sqrt[n]{l_2/l_{r+1}}$ define a local covering of degree n^2 at $E \cap D_1$ and local coordinates in Y. Here, E is the exceptional divisor of the blown-up singular point. We can treat the other $E \cap D_i$ in the same way.

The function field of Y is given by

$$\mathbb{C}\left(\frac{x_1}{x_0}, \frac{x_2}{x_0}, \sqrt[n]{l_i/l_j}\right), \quad i, j = 1, \ldots, k,$$

where $(x_0 : x_1 : x_2)$ are homogeneous coordinates of $\mathbb{P}_2$. It is a Kummer extension of degree n^{k-1} of the function field of X, which is $\mathbb{C}(x_1/x_0, x_2/x_0)$, the function field of $\mathbb{P}_2$. In general, we call Y a *Kummer covering* of X if the function field of Y is a Kummer extension of the function field of X. That is, the function field of Y is obtained by adjoining to the function field of X roots of that field. The surface $Y = Y(C, n)$ satisfies

$$n^2 \frac{3c_2 - c_1^2}{N}(Y) = (n-1)^2(f_0 - k) - 2(n-1)(f_1 - 2f_0) + 4(f_0 - t_2),$$

$$(B.1)$$

where

$$f_0 = \sum_{r \geq 2} t_r$$

and

$$f_1 = \sum_{r \geq 2} r t_r.$$

For the proof, we use formula (3.12) of Chapter 3, adapted to our situation:

$$\frac{n^2}{N}(3c_2 - c_1^2)(Y) = n^2(3c_2(X) - c_1^2(X)) + n(n-1)\sum_i(-e(D_i') + 2(D_i')^2)$$

$$+(n-1)^2 \sum_{i<j} D_i' \cdot D_j' - (n-1)^2 \sum_i (D_i')^2.$$

The D_i' in this formula denote the D_i together with the E_j of the previous discussion. The surface X is $\mathbb{P}_2$ with $f_0 - t_2$ points blown up. Therefore,

$$3c_2(X) - c_1^2(X) = 4(f_0 - t_2).$$

Moreover, we have

$$\sum_i e(D_i') = 2(k + f_0 - t_2)$$

$$\sum_i (D_i')^2 = \sum_{j=1}^{k}(1 - \sigma_j) - (f_0 - t_2)$$

$$= k - (f_1 - 2t_2) - (f_0 - t_2)$$

$$= k - f_0 - f_1 + 3t_2,$$

where σ_j is the number of singular points on the jth line. Finally, we have

$$\sum_{i<j} D_i' \cdot D_j' = f_1 - t_2,$$

and the desired formula (B.1) for the quadratic polynomial in $n - 1$ follows. Here, we have repeated some of the arguments used in Chapter 5, §5.2, to derive Höfer's formula (5.11). Recall that formula:

$$\mathrm{Prop}(Y)/N = \frac{1}{4}\sum_j P_j^2 + \frac{1}{4}\sum_{i,l} R_{il} x_i x_l.$$

If we construct a Kummer covering from an arrangement with $3\tau_i = k + 3$ for all lines, then the second term of the right-hand side of this equation vanishes (see Chapter 5, §5.3). In this case, the quadratic polynomial is

$$n^2 \mathrm{Prop}(Y)/N = \frac{1}{4}\sum_{r \geq 3} t_r (n(r - 2) - (r + 2))^2,$$

which is a nice example of the Miyaoka-Yau inequality. The number $\mathrm{Prop}(Y)$ vanishes exactly when all but one of the t_r ($r \geq 3$) vanish and, for the non-vanishing t_r, we have $n = \frac{r+2}{r-2}$. This diophantine condition has the solutions

$$r = 3, \; n = 5; \; r = 4, \; n = 3; \; r = 6, \; n = 2.$$

No configuration is known with $3\tau_i = k + 3$ and $t_r = 0$ for $r \neq 2$, 6. However, in the cases where t_r vanishes for $r \neq 2$, 3 and $r \neq 2$, 4, respectively, we have the following examples:

I) The complete quadrilateral with $k = 6$, $t_2 = 3$, $t_3 = 4$ and

$$n^2 \mathrm{Prop}(Y)/N = (n - 5)^2.$$

II) The Hesse arrangement with $k = 12$, $t_2 = 12$, $t_4 = 9$ and

$$n^2 \mathrm{Prop}(Y)/N = 9(n - 3)^2.$$

III) The arrangement Ceva(3) with $k = 9$, $t_2 = 0$, $t_3 = 12$ and

$$n^2 \mathrm{Prop}(Y)/N = 3(n - 5)^2.$$

The reader may check that we also obtain these quadratic polynomials by applying (B.1) above. In the cases I), II), III) we obtain ball quotients Y for $n = 5$, 3, 5, respectively. These surfaces Y are certainly of general type, since $c_1^2(Y)$ is greater than 9. Namely, $c_1^2(Y)$ equals $3^2 \cdot 5^4$, $2^4 \cdot 3^{11}$, $3^2 \cdot 5^6 \cdot 37$, respectively, as can be calculated using the formulas of Chapter 3. Notice that case I) corresponds to the complete quadrilateral with the covering for which $\mu_j = \frac{2}{5}$ (see Chapter 3, §3.3).

For every arrangement C we have the surface $Y = Y(C, n)$. According to the discussion in Chapter 4, we know that $\mathrm{Prop}(Y(C, n)) \geq 0$ if $Y(C, n)$ is not in the class given by 3) of Theorem ROC, Chapter 4, §4.1. In particular, if we also have $n = 3$, then, by (B.1),

$$f_0 - k - (f_1 - 2f_0) + f_0 - t_2 \geq 0$$

or, equivalently,

$$t_2 + t_3 \geq k + \sum_{r \geq 4} (r - 4)t_r. \tag{B.2}$$

If $k > 4$ and $t_{k-1} = 1$, then we have $t_2 = k - 1$, $t_3 = 0$, and the inequality in (B.2) does not hold. Of course, in this case $Y(C, 3)$ belongs to class 3) of Theorem ROC (Theorem 4.1), §4.1. It is a ruled surface over a curve of genus at least 2 (the fibers coming from the lines passing through the $(k - 1)$-fold point). We have the following.

Theorem B.1 *If $t_k = t_{k-1} = t_{k-2} = 0$ and if $k \geq 6$, then $Y(C, n)$ is of general type for $n \geq 3$.*

As part of the proof of this result, we construct a nonnegative double canonical divisor on X. Then, we know the classes 1), 2), 3) of Theorem ROC are excluded. We use formal canonical divisors with half-integral coefficients. Now, by the assumption of the proposition, we can find six lines $L_1, L_2, \ldots, L_6$ of the arrangement such that no more than three of them pass through one point. For K on $\mathbb{P}_2$ we take $-\frac{1}{2}(L_1 + \ldots + L_6)$. Let τ be the projection of X onto $\mathbb{P}_2$. Consider the divisor, with coefficients in $\mathbb{Q}$,

$$\widetilde{K}_Y = -\frac{1}{2}\tau^*(L_1 + \ldots + L_6) + \sum_j E_j + \left(1 - \frac{1}{n}\right)\left(\sum_j E_j + \sum_i D_i\right)$$

on X. Its lift to Y is a canonical divisor that is nonnegative for $n \geq 2$, because E_j has multiplicity at least $-\frac{3}{2}$ in $-\frac{1}{2}\tau^*(L_1 + \ldots + L_6)$. For the complete proof that $Y(C, n)$ is of general type for $n \geq 3$, we refer you to [11]. In any case, Theorem B.1 implies the inequality (B.2) for $k \geq 6$ and $t_k = t_{k-1} = t_{k-2} = 0$.

There are several refinements of the inequality (B.2). Checking the remaining cases, we obtain the surprising result:

Theorem B.2 *Given an arrangement with $t_k = 0$, in other words the arrangement is not a pencil, either t_2 or t_3 is different from 0.*

This simple geometric fact has amusing consequences. But, up to now, it has only been proved using the Miyaoka-Yau inequality (see [11]).

The above result is incorrect for finite fields with more than two elements. Namely, consider the projective plane over the field $\mathbb{F}_q$ of q elements. Then, for the arrangement of *all lines*, we have

$$k = t_{q+1} = q^2 + q + 1,$$

with all other $t_r = 0$.

For *real* arrangements, Sylvester (1893) asked whether $t_2 \neq 0$ if the arrangement is not a pencil. This, for complex arrangements, is certainly incorrect (see Ceva(3)). In 1940, Melchior utilized the cellular decomposition of the real projective plane defined by the arrangement to answer Sylvester's question in the affirmative. Let f_0 be the number of vertices in the cellular decomposition. Then,

$$f_0 = \sum_{r \geq 2} t_r$$

and, if p_r is the number of r-gons, we have that the number f_1 of edges is given by

$$f_1 = \sum_{r \geq 2} r\, t_r = \frac{1}{2} \sum_{r \geq 3} r\, p_r$$

and

$$f_2 = \sum_{r \geq 3} p_r.$$

Notice that $p_2 = 0$, as we do not have a pencil. Now, as $e(\mathbb{P}_2(\mathbb{R})) = 1$, we have

$$3 - 3(f_0 - f_1 + f_2) = 0,$$

which is equivalent to

$$3 + \sum_{r \geq 2} (r - 3) t_r + \sum_{r \geq 3} (r - 3) p_r = 0.$$

Hence,
$$t_2 \geq 3.$$

A line arrangement determines a triangulation if and only if it satisfies

$$t_2 = 3 + \sum_{r \geq 4} (r - 3) t_r.$$

It is conjectured that t_2 is at least $\frac{k}{2}$ if the arrangement is not a pencil and if $k \neq 7$ and $k \neq 13$. In those cases, there are arrangements with $t_2 = 3$ and $t_2 = 6$, respectively. For the history of this problem, see [32].

A Kummer covering for the icosahedral arrangement

Recall that the icosahedral arrangement is a real arrangement defined as follows. The icosahedron can be inscribed in the sphere $S^2 \subset \mathbb{R}^3$, with all vertices on the sphere. It can then be projected onto the sphere from the origin. The thirty edges, forming fifteen pairs of antipodal edges, define fifteen great circles on S^2. Passing from S^2 to $\mathbb{P}_2(\mathbb{R})$ by identifying antipodal points, the fifteen great circles define fifteen projective lines forming an arrangement with $k = 15$, $t_2 = 15$, $t_3 = 10$, $t_5 = 6$, and $t_r = 0$ otherwise. The double points correspond to the midpoints of the thirty edges, the triple points to the midpoints of the twenty faces, and the quintuple points to the twelve vertices. The arrangement can be drawn as follows. The six quintuple points are called poles. We distinguish one pole and take it as the center of a regular pentagon whose vertices are the five other poles. The fifteen lines of the arrangement are the connecting lines of the six poles. The fifteen lines can be divided into five sets of three lines, with the lines in any one set meeting in double points. These five "triangles" correspond to the five cubes in which the icosahedron can be inscribed. The symmetry group of the icosahedron is a subgroup of $SO(3)$; it is isomorphic to the alternating group A_5 operating on the five cubes.

Now, consider the icosahedral arrangement of lines in the complex projective plane. We now construct a Kummer covering of X, where X is $\mathbb{P}_2$ with the six quintuple points and the ten triple points blown up. We assign the ramification 5 to each of the fifteen lines and also to the exceptional divisors defined by the triple points. We assign ramification 1 to the exceptional divisors defined by the quintuple points, which are the poles. The icosahedral arrangement satisfies $3\tau = k + 3 = 18$, because on each line there are two double points, two triple points and two quintuple points. A good covering Y of X of degree N, with ramifications as above, satisfies

$$3c_2(Y) - c_1^2(Y) = 0,$$

because

$$\frac{2}{m} + \frac{r}{n} = r - 2$$

(see Chapter 5, §5.4) holds for $m = 5$, $n = 5$, $r = 3$ and $m = 1$, $n = 5$, $r = 5$. Before constructing Y, we shall calculate the signature of Y using formula (3.18) of Chapter 3. We have

$$3\text{sign}(Y)/N = 3(-15) - 15\frac{24}{25}(-3) - 10\frac{24}{25}(-1) = \frac{39}{5}.$$

As $\text{Prop}(Y)/N = 0$, we have

$$c_2(Y)/N = \frac{39}{5}, \quad c_1^2(Y)/N = \frac{3 \times 39}{5}.$$

The covering Y we construct has $N = 25$. Hence $c_2(Y) = 195$ and $c_1^2(Y) = 585$. Therefore, Y will be a ball quotient of general type (see Theorem ROC, §4.1). To construct Y as a Kummer covering with $N = 25$, we proceed as follows. The distinguished pole is the center of the regular pentagon whose vertices we number cyclically by $P_1, P_2, \ldots, P_5$. We have

- five lines joining the P_i to the distinguished pole, represented by an equation $A = 0$,
- five lines joining P_i to P_{i+1}, represented by an equation $B = 0$,
- five lines joining P_i to P_{i+2}, represented by an equation $C = 0$.

Let $l_0 = 0$ be the equation of an arbitrary line not in the arrangement. Then,

$$\sqrt[5]{A^3 B / l_0^{20}}, \quad \sqrt[5]{A^3 C / l_0^{20}}$$

define the Kummer covering of $\mathbb{P}_2$, and of X, of degree 25. The five lines corresponding to $A = 0$, $B = 0$, $C = 0$ will be denoted by A_i, B_i, C_i respectively. Then, we check the following. Through the distinguished pole pass five lines A_i; through the others pass one line A_i, two lines B_i, and two lines C_i. Through each of the threefold points pass one line A_i and two lines B_i, or one line A_i and two lines C_i. It is therefore easy to see that we obtain a cover of X with the ramifications prescribed above, which is a good covering, that is, it satisfies conditions (i), (ii), and (iii) of Definition 5.1. Notice that this Kummer covering is of a different type than that considered at the beginning of the chapter. For the icosahedral arrangement, that construction would not yield a ball quotient.

Bibliography

[1] R. C. Alperin, *An elementary account of Selberg's Lemma*, L' Enseignement Math., 33 (1987), 269–273.

[2] M. T. Anderson, *A survey of the Einstein metrics on 4-manifolds*, in Handbook of Geometric Analysis, No. 3, Adv. Lectures Math., Vol. XIV, eds. L. Ji, P. Li, R. Schoen, L. Simon, 2010, 1–35

[3] P. L. Antonelli, Handbook of Finsler Geometry, Volume 1, Kluwer Academic Pub., 2003.

[4] P. Appell, *Sur les séries hypergéométriques de deux variables et sur les équations différentielles linéaires aux dérivées partielles*, C. R. Acad. Sci. Paris, 90 (1880) 296–298, 731–735.

[5] P. Appell, *Sur les fonctions hypergéométriques de deux variables*, J. Math. Pures Appl. (3) 8 (1882), 173–216.

[6] P. Appell, M. J. Kampé de Fériet, *Fonctions hypergéométriques et hypersphériques, Polynômes d'Hermite*, Gauthier–Villars, Paris, 1926.

[7] T. Aubin, *Equations du type de Monge-Ampère sur les variétés kähleriennes compactes*, C. R. Acad. Sci. Paris, 283 (1976), 119–121.

[8] L. Auslander, *Bieberbach's theorems on space groups and discrete uniform subgroups of Lie groups*, Annals of Math., 71, No. 3 (1960) 579–590.

[9] S. Axler, P. Bourdon, W. Ramey, Harmonic Function Theory, 2nd. ed., GTM 137, Springer 2010.

[10] L. Badescu, Algebraic Surfaces, Universitext, Springer-Verlag New York Inc. 2001.

[11] G. Barthel, F. Hirzebruch, Th. Höfer, Geradenkonfigurationen und Algebraische Flächen, Vieweg (1987).

[12] W. Barth, C. Peters, A. Van de Ven, Compact Complex Surfaces, Ergebnisse, 3 Folge Band 4, Springer 1984.

[13] A. Beauville, Complex Algebraic Surfaces, Second Edition, LMS Student Texts 34, 1996.

[14] M. Berger, B. Gostiaux, Differential Geometry: Manifolds, Curves and Surfaces, GTM 115, Springer-Verlag (1988) (translation of original 1987 French version, PUF).

[15] L. Bieberbach, *Über die Bewegungsgruppen der Euklidischen Räume*, Math. Ann. 70 (1911), 297–336.

[16] E. Bombieri, *The Pluricanonical Map of a Complex Surface*, in Several Complex Variables 1, Maryland 1970, LNM 155 (1970), 35–87.

[17] A. Borel, *Compact Clifford-Klein forms of symmetric spaces*, Topology 2 (1963), 111–122.

[18] A. Borel, Introduction aux groupes arithmétiques, Hermann 1969.

[19] R. Bott, L. W. Tu, Differential Forms in Algebraic Toplogy, GTM 82, Springer 1982.

[20] J.-P. Bourguignon, *Premières formes de Chern des Variétés Kählériennes Compactes (d'après E. Calabi, T. Aubin et S.T. Yau)*, Séminaire Bourbaki, 30e année, 507 (1977–78), 1–21.

[21] C. Boyer, K. Galicki, Sasakian Geometry, Oxford Math. Monographs, Oxford 2008.

[22] J. L. Brenner, R. C. Lyndon, *A theorem of G.A. Miller on the order of the product of two permutations; I*, Jñānābha, 14 (1984), 1–16.

[23] E. Brieskorn, *Rationale Singularitäten Komplexer Flächen*, Inventiones Math. 4 (1968), 336–358.

[24] S. Bundgaard, J. Nielsen, *On normal subgroups with finite index in F-groups*, Mat. Tidsskrift B (1951), 56–58.

[25] C. Carathéodory, Theory of Functions II, Chelsea, (1954) (Funktionentheorie II, Birkhäuser, Basel, 1950).

[26] H. Cartan, *Quotient d'un espace analytique par un groupe d'automorphismes*, in Algebraic Geometry and Topology, Symposium in Honor of S. Lefschetz, Princeton University Press, 1957, 90–120.

[27] S. Y. Cheng, S. T. Yau, *Inequality between Chern numbers of singular Kähler surfaces and characterization of orbit space of discrete group of SU(2,1)*, Contemp. Math. 49 (1986), 31–45.

[28] C. Chevalley, *Invariants of finite groups generated by reflections*, American J. Math. 77 (1955), 778–782.

[29] P. B. Cohen, F. Hirzebruch, Book Review: Commensurabilities among lattices in PU(1, n), by P. Deligne and G. D. Mostow, Annals of Math. Studies, no. 132, Bull. American Math. Soc., 32, No. 1, Jan. 1995, 88–105.

[30] P. B. Cohen, J. Wolfart, *Fonctions hypergéométriques en plusieurs variables et espaces des modules de variétés abéliennes*, Ann. Sci. Ecole Norm. Sup. (4), 26 (1993), 665–690.

[31] H. Cohn, Conformal Mapping on Riemann Surfaces, Dover Books on Mathematics, 2010. First published in 1967 by McGraw-Hill.

[32] J. Csima, E. T. Sawyer, *There Exist 6n/13 Ordinary Points*, Discrete Comp. Geom. 9 (1993), 187–202.

[33] P. Deligne, G. D. Mostow, *Monodromy of Hypergeometric functions*, Publ. Math. IHES, 63 (1986), 5–90.

[34] P. Deligne, G. D. Mostow, Commensurabilities among lattices in PU(1, n), Annals of Math. Studies, no. 132, Princeton Uni. Press, Princeton, NJ (1993).

[35] M. Demazure, *A, B, C, D, E, F, etc.* in Sémin, sur les singularités de surfaces, LNM 777, Springer, Heidelberg (1980), 222–227.

[36] A. L. Edmonds, J. H. Ewing, R. S. Kulkarni, *Regular tessellations of surfaces and* $(p, q, 2)$-*triangle groups*, Ann. of Math., 116 (1982), 113–132.

[37] A. L. Edmonds, J. H. Ewing, R. S. Kulkarni, *Torsion free subgroups of Fuchsian groups and tessellations of surfaces*, Invent. Math., 69 (1982), 331–346.

[38] A. Erdélyi, *The analytic theory of systems of partial differential equations*, Bill. Amer. Math. Soc. 57 (1951), 339–353.

[39] H. M. Farkas, I. Kra, Riemann Surfaces, GTM 71, Second Edition, Springer-Verlag (1992).

[40] W. Fenchel, *Bemarkingen om endelige gruppen af abbildungsklasser*, Mat. Tids-skrift B (1950), 90–95.

[41] R. D. Feuer, *Torsion-free subgroups of triangle groups*, Proc. Amer. Math. Soc., 30 (1971), 235–240.

[42] R. H. Fox, *On Fenchel's conjecture about F-groups*, Mat. Tidsskrift B (1952), 61–65.

[43] R. H. Fox, *Covering spaces with singularities*, in Algebraic Geometry and Topology, Princeton Univ. Press (1957), 243–257.

[44] R. Friedman, Algebraic surfaces and holomorphic vector bundles, Springer-Verlag (1998).

[45] L. Fuchs, *Zur Theorie der linearen Differentialgleichungen mit veränderlichen Coeffizienten*, J. Reine Angew. Math., 66 (1866), 121–162.

[46] A. Ghigi, J. Kollár, *Kähler-Einstein metrics on orbifolds and Einstein metrics on spheres*, Comment. Math. Helv., 82(4) (2007), 877–902.

[47] E. Goursat, Leçons sur les séries hypergéométriques et sur quelques fonctions qui s'y rattachent, Actualités scientifiques et industrielles 333, Hermann 1936.

[48] H. Grauert, *Über Modifikationen und exzeptionelle analytische Mengen*, Math. Ann. 146 (1962), 331–368.

[49] P. Griffiths, J. Harris, Principles of Algebraic Geometry, John Wiley and Sons, 1978; Wiley Classics Library, 1994.

[50] H. Guggenheimer, *Über vierdimensionale Einsteinräume*, Experientia 8 (1952), 420–421.

[51] R. Hartshorne, Algebraic Geometry, 6th edition, New York, Springer-Verlag (1993) c1977, Graduate Texts in Mathematics **52**.

[52] J. C. Hemperley, *The parabolic contribution to the number of linearly independent automorphic forms on a certain bounded domain*, Amer. J. Math. 94 (1972), 1028–1100.

[53] F. Hirzebruch, *Über eine Klasse von einfach-zusammenhängenden komplexen Mannigfaltigkeiten*, Math. Ann. 124 (1951), 77–86.

[54] F. Hirzebruch, *Automorphe Formen und der Satz von Riemann-Roch*, In: Symposium Internacional de Topologia Algebraica (México 1956), Univ. Nacional. Autónoma de Mexico (1958), 129–144.

[55] F. Hirzebruch, Neue topologische Methoden in der algebraischen Geometrie, Erg. Math. u.i. Grenz., Neue Folge, Heft 9, Berlin, Springer-Verlag (1956). Topological methods in algebraic geometry, Grund. der math. Wissensch., Bd 131 (1966), reprinted "Classics in Mathematics," Springer (1995).

[56] F. Hirzebruch, *The signature theorem: Reminiscences and recreation*, In: Prospects in Mathematics, Ann. Math. Stud., 70 (1971), 3–31.

[57] F. Hirzebruch, *Singularities of algebraic surfaces and characteristic classes*, Contemp. Math. 58, I (1986), 141–155.

[58] F. Hirzebruch, A. Van de Ven, *Hilbert modular surfaces and the classification of algebraic surfaces*, Invent. Math., 23 (1974), 1–29.

[59] Th. Höfer, Ballquotienten als verzweigte Überlagerungen der projective Ebene, Ph.D. thesis. Bonn 1985.

[60] R.-P. Holzapfel, *A class of minimal surfaces in unknown region of surface geography*, Math. Nachr. 98 (1980), 221–232.

[61] R.-P. Holzapfel, *Arithmetic surfaces with great K^2*, Proc. Alg. Geom. Bucharest 1980, Teubner, Leipzih 1981, 80–91.

[62] R.-P. Holzapfel, *Invariants of arithmetic ball quotient surfaces*, Math. Nachr. 103 (1981), 117–153.

[63] R.-P. Holzapfel, *Chern numbers of algebraic surfaces—Hirzebruch's examples are Picard modular surfaces*, Math. Nachr. 126 (1986), 255–272.

[64] R.-P. Holzapfel, *Basic two-dimensional versions of Hurwitz genus formula*, Ann. Global Anal. Geom. 4 (1986), 1–70.

[65] H. Hopf, Schlichte Abbildungen und lokale Modifikationen 4 dimensionaler komplexer Mannigfaltigkeiten, Comm. Math. Helv. 29 (1955), 132–156; in Selecta Springer-Verlag, 1964, 284–306.

[66] L. Hörmander, An Introduction to Complex Analysis in Several Variables, 3rd ed. (revised), North-Holland Math. Library (1990).

[67] C. Horst, S. Kobayashi, H.-H. Wu, Complex Differential Geometry, Birkhäuser, Basel and Boston, (1983).

[68] S. Iitaka, *Logarithmic forms of algebraic varieties*, Jour. Fac. Science Univv. Tokyo, Sec. 1A, Math. 23 (3) (1976), 525–544.

[69] S. Iitaka, *Geometry on complements of lines in $\mathbb{P}^2$*, Tokyo J. Math. 1 (1978), 1–19.

[70] S. Iitaka, Algebraic Geometry, Springer-Verlag (1982).

[71] K. Jänich, Topology, UTM, Springer-Verlag (1984).

[72] D. D. Joyce, Compact Manifolds with Special Holonomy, Oxford Mathematical Monographs, Oxford Science Pub., 2000.

[73] E. Kähler, *Über eine bemerkenswerte Hermitesche Metrik*, Abh. Math. Sem. Hamburg Univ. 9 (1933), 173–186.

[74] M. Kato, *On the existence of finite principal uniformization of $\mathbb{C}\mathbb{P}^2$ along weighted configurations*, Mem. Fac. Sci. Kyushu Univ. Ser. A, 38 (1984), 127–131.

[75] M. Kato, *On uniformizations of orbifolds*, Advanced Studies in Pure Math., 9 (1986), 149–172.

[76] F. Klein, *Über die Transformation siebenter Ordnung der elliptischen Funktionen*, Math. Ann. 14 (1878), Ges. Math. Abh. Band 3, LXXXIV, 90–136, Nachdruck, Springer 1973.

[77] A. W. Knapp, *Doubly generated Fuchsian groups*, Michigan Math. J. 15 (1968), 289–304.

[78] L. Kaup, B. Kaup, Holomorphic Functions of Several Variables, de Gruyter, Berlin (1983).

[79] R. Kobayashi, *Kähler-Einstein metrics on an open algebraic manifold*, Osaka J. Math. 21 (1984), 399–418.

[80] R. Kobayashi, *A remark on the Ricci curvature of algebraic surfaces of general type*, Tôhoku Math. J. 36 (1984), 385–399.

[81] R. Kobayashi, *Einstein-Kähler metrics on open algebraic surfaces of general type*, Tôhoku Math. J., 37 (1985), 43–77.

[82] R. Kobayashi, *Einstein-Kähler V-metrics on open Satake V-surfaces with isolated singularities*, Math. Ann. 272 (1985), 385–398.

[83] R. Kobayashi, S. Nakamura, F. Sakai, *A numerical characterization of ball quotients for normal surfaces with branch loci*, Proceedings if the Japan Academy 65, Ser. A, No. 7 (1989), 238–241.

[84] R. Kobayashi, *Uniformization of complex surfaces*, Advanced Studies in Pure Math., 18-II, Kähler Metrics and Moduli Spaces, 313–394, 1990.

[85] S. Kobayashi, K. Nomizu, Foundations of Differential Geometry I,II, Wiley, New York 1963, 1969.

[86] K. Kodaira. *On Kähler varieties of restricted type (an intrinsic characterization of algebraic varieties)*, Annals of Math., 60 (1954), 28–48.

[87] K. Kodaira, *On the structure of complex analytic spaces IV*, American J. of Math., 90 (1968), 1048–1066.

[88] K. Kodaira, *Pluricanonical systems on algebraic surfaces of general type*, J. Math. Soc. Japan, 20 (1968), 170–192.

[89] I. Kra, Automorphic Forms and Kleinian Groups, Benjamin, Massachusetts (1972).

[90] S. Lang, Elliptic Functions, Addison-Wesley, 1973.

[91] A. Langer, *The Bogomolov-Miyaoka-Yau inequality for log canonical surfaces*, J. London Math. Soc. 64 (2001), 327–343.

[92] A. Langer, *Logarithmic orbifold Euler numbers with applications*, Proc. London Math. Soc. (3) 86 (2003), 358–396.

[93] G. Lauricella, *Sulle funzioni ipergeometriche a piu variabli*, Rend. Circ. Mat. Palermo 7 (1893), 111–158.

[94] R. C. Lyndon, Groups and Geometry, L. M. S. Lecture Note Series 101, Cambridge University Press 1985.

[95] R. C. Lyndon, P.E. Schupp, Combinatorial Group Theory, Ergebnisse der Mathematik und ihrer Grenzgebiete 89, Springer-Verlag 1977.

[96] A. M. Macbeath, Fuchsian Groups, Queen's College Dundee 1961.

[97] W. Magnus, Noneuclidean Tesselations and Their Groups, Pure and Applied Mathematics 61, Academic Press 1974.

[98] B. Maskit, Kleinian Groups, Springer-Verlag (1988).

[99] K. H. Mayer, Algebraische Topologie, Birkhäuser (1989).

[100] J. Mennicke, *Eine Bemerkung über Fuchssche Gruppen*, Invent. Math., 2 (1967), 301–305. Corrigendum: Invent. Math., 6 (1968), 106.

[101] G. A. Margulis, Discrete subgroups of semisimple Lie groups, Ergeb. der Math. 17, Springer–Verlag 1991.

[102] G. A. Miller, *On the product of two substitutions*, Amer. J. Math., 22 (1900), 185–190.

[103] J. Milnor, *On the 3-dimensional Brieskorn manifolds $M(p,q,r)$*, in: Knots, Groups, and 3-Manifolds, ed. by L. P. Neuwirth. Annals of Mathematics Studies 84, Princeton University Press 1975.

[104] Y. Miyaoka, *On the Chern numbers of surfaces of general type*, Invent. Math. 42 (1977), 225–237.

[105] Y. Miyaoka, *The maximal number of quotient singularities on surfaces with given numerical invariants*, Math. Ann. 268 (1984), 159–171.

[106] S. Mochizuki, *Correspondences on hyperbolic curves*, J. Pure and Applied Algebra, 131 (1998), 227–244.

[107] B. G. Moishezon, *A criterion for projectivity of complete algebraic abstract varieties*, Amer. Math. Soc. Translations, 63 (1967), 1–50.

[108] G. D. Mostow, *Generalized Picard lattices arising from half-integral conditions*, Publ. Math. IHES, 93 (1986), 91–106.

[109] G. D. Mostow, *On discontinuous action of monodromy groups on the complex n-ball*, J. Amer. Math. Soc. 1 (1988), 171–276.

[110] D. Mumford, *The canonical ring of an algebraic surface*, Annals Math. 76 (1962), 612–615.

[111] D. Mumford, *The topology of normal singularities of an algebraic surface and a criterion for simplicity*, Inst. Hautes Études Sci. Publ. Math. 9 (1961), 5–22.

[112] J. R. Munkres, Elements of Algebraic Topology, Addison-Wesley, 1984.

[113] Y. Nakai, *A criterion of an ample sheaf on a projective scheme*, Amer. J. Math., 85 (1963), 14–26.

[114] S. Nakamura, *Classification and uniformization of log-canonical surface singularities in the presence of branch loci (in Japanese)*, Master's thesis, Saitama Univ. 1989.

[115] J. Nielsen, *Kommutatorgruppen for det frie produkt af cykliske grupper*, Mat. Tidsskrift B (1948), 49–56.

[116] E. Picard, *Sur une extension aux fonctions de deux variables du problème de Riemann relatif aux fonctions hypergéométriques*, Ann. Sci. ENS, 10 (1881), 305–322.

[117] T. Radó, *Über den Begriff der Riemannschen Fläche*, Acta Szeged, 2 (1925), 101–121.

[118] R. Remmert, Theory of complex functions; translated by Robert B. Burckel, GTM 122, Springer-Verlag New York, 1991.

[119] B. Riemann, *Beiträge zur Theorie der durch die Gauss'sche Reihe $F(\alpha, \beta, \gamma, x)$ darstellbaren Functionen*, Abh. Kön. Ges. Wiss. Göttingen Math Cl. VII (1857).

[120] J. Ross, R. Thomas, *Weighted projective embeddings, stability of orbifolds and constant scalar curvature Kähler metrics*, Jour. Diff. Geom. 88 (2011), p.109–160.

[121] C.-H. Sah, *Groups related to compact Riemann surfaces*, Acta math., 123 (1969), 13–42.

[122] F. Sakai, *Semi-stable curves on algebraic surfaces and logarithmic pluricanonical maps*, Math. Ann. 254 (1980), 89–120.

[123] F. Sakai, *Weil divisors on normal surfaces*, Duke Math. J., 51 (1984), 877–887.

[124] F. Sakai, *Classification of normal surfaces*, Proc. Symp. Pure Math., 46 (1987), 451–465.

[125] I. Satake, *Om a generalization of the notion of manifold*, Proc. Nat. Acad. Sci. USA, 42 (1956), 359–363.

[126] I. Satake, Algebraic Structures of Complex Symmetric Domains, Princeton Math. Series, Iwanami Shoten and Princeton University Press, 1980.

[127] J. K. Sauter, *Isomorphisms among monodromy groups and applications to lattices of $PU(1, 2)$*, Pacific J. Math. 146 (1990), 331–384.

[128] H. A. Schwarz, *Ueber diejenigen Fälle, in welchen die Gauss'sch hypergeometrishce Reihe eine algebraische Function ihres vierten Elements darstellt*, J. Reine Angew. Math. 75 (1873), 292–335.

[129] A. Selberg, *On discontinuous groups in higher dimensional symmetric spaces*, in: Contributions to Function Theory, Tata Institute of Fundamental Research, Bombay 1960.

[130] I. R. Shafarevich, Basic Algebraic Geometry 1, Varieties in Projective Space, Second Edition, Springer-Verlag Berlin, Heidelberg, 1994.

[131] I. R. Shafarevich, Basic Algebraic Geometry 2, Schemes and Complex Manifolds, Second Edition, Springer-Verlag Berlin, Heidelberg, 1997.

[132] G. C. Shephard, J. A. Todd, *Finite unitary reflection groups*, Canadian J. Math. 6 (1954), 274–304.

[133] V. V. Shokurov, *Riemann surfaces and Algebraic Curves*, in: Algebraic Geometry, Encyclopedia of Mathematical Sciences, Vol 23, I.R. Shafarevich (Ed.).

[134] K. Takeuchi, *Arithmetic triangle groups*, J. Math. Soc. Japan, 29 (1977), 91–106.

[135] K. Takeuchi, *Commensurability classes of arithmetic discrete triangle groups*, J. Fac Sci. Univ. Tokyo, 24 (1977), 201–212.

[136] T. Terada, *Problème de Riemann et fonctions automorphes provenant des fonctions hypergéométriques de plusieurs variables*, J. Math. Kyoto. Univ., 13 (1973), 557–578.

[137] W. Thurston, The geometry and topology of three-manifolds, Princeton Univ. (mimeographed notes 1978–1979), updated at http://library.msri.org/books/gt3m/PDF/1.pdf.

[138] A. M. Uludağ, *Orbifolds and their uniformization*, in Arithmetic and Geometry around Hypergeometric Functions, Progress in Math. Vol. 260, 373–406, Birkhäuser, 2007.

[139] A. Van de Ven, *On the Chern numbers of surfaces of general type*, Invent. Math. 36 (1976), 285–293.

[140] A. Van de Ven, *Some recent results on surfaces of general type*, Sem. Bourbaki 1977, Exp. 5000, Springer LNM 677 (1978), 155–166.

[141] C. Vuono, *The Kodaira Embedding Theorem for Kähler Varieties with Isolated Singularities*, J. Geometric Analysis 3, no.5 (1993), 403–421.

[142] H. Weyl, Die Idee der Riemannschen Fläche, B. G. Teubner, Leipzig, Berlin, (1913), reedition (ed. R. Remmert), B. G. Teubner (1997).

[143] S. T. Yau, *Calabi's conjecture and some new results in algebraic geometry*, Proc. Nat. Acad. Sci. USA 74 (1977), 1798–1799.

[144] S. T. Yau, *On the Ricci curvature of a compact Kähler manifold and the complex Monge-Ampère equation, I*, Pure Appl. Math. 31 (1978), 339–411.

[145] M. Yoshida, Fuchsian differential equations, Aspects of Mathematics, Vieweg, (1987).

[146] H. Zieschang, E. Vogt, H.-D. Coldewey, Flächen und ebene diskontinuierliche Gruppen, Lecture Notes in Mathematics 122, Springer-Verlag 1970.

[147] H. Zieschang, E. Vogt, H.-D. Coldewey, Surfaces and planar discontinuous groups, Lecture Notes in Mathematics 835, Springer-Verlag 1980.

Index